Augmented Reality

This book focuses on augmented reality (AR) technology, which uses the real environment to superimpose virtual elements. Therefore, the reader can create applications that simulate scenarios that can be dangerous or expensive to generate in the real world. AR has proven helpful in education, marketing, and industrial scenarios. AR technology improves the user experience of various disciplines, incorporating virtual information that maximizes the experience and adds knowledge. This book intends students, researchers, and developers to have the possibility of finding the foundations on which AR technology rests.

Our book intends that students, researchers, and developers: (i) learn the basics of AR; (ii) understand the technologies that support AR; (iii) know about AR applications that have been a watershed; (iv) gain an understanding of the critical elements needed to implement an AR application; (v) acquire skills in the step-by-step development of an AR application; (vi) learn how to use the instruments to evaluate an AR application; (vii) understand how to present the information about study cases; and (viii) gain knowledge about AR challenges and trends.

Augmented Reality

Fundamentals and Applications

Osslan Osiris Vergara Villegas and
Vianey Guadalupe Cruz Sánchez

CRC Press
Taylor & Francis Group
Boca Raton London New York

CRC Press is an imprint of the
Taylor & Francis Group, an **Informa** business

Cover image credit: Itziar Vergara

First edition published 2024
by CRC Press
2385 NW Executive Center Drive, Suite 320, Boca Raton FL 33431

and by CRC Press
4 Park Square, Milton Park, Abingdon, Oxon, OX14 4RN

CRC Press is an imprint of Taylor & Francis Group, LLC

ISBN: 978-1-032-56371-8 (hbk)
ISBN: 978-1-032-56372-5 (pbk)
ISBN: 978-1-003-43519-8 (ebk)

DOI: 10.1201/9781003435198

Typeset in Times
by Newgen Publishing UK

Dedication

To our dear children Itziar and Osslan.
Everything happens thanks to you.
To our parents.

Contents

About the Authors

Osslan Osiris Vergara Villegas was born in Cuernavaca, Morelos, Mexico, on July 3, 1977. He earned a B.Sc. degree in Computer Engineering from the Instituto Tecnologico de Zacatepec, Mexico, in 2000; an M.Sc. in Computer Science at the Center of Research and Technological Development (CENIDET) in 2003; and a Ph.D. degree in Computer Science from CENIDET in 2006. He is a Professor at the Universidad Autónoma de Ciudad Juárez, Chihuahua, Mexico, where he heads the Computer Vision and Augmented Reality laboratory. Prof. Vergara is a level-one member of the Mexican National Research System. He serves several peer-reviewed international journals and conferences as an editorial board member and reviewer. He has co-authored over 100 book chapters, journals, and international conference papers. Vergara has directed more than 50 B.S., M.Sc., and Ph.D. theses. He is a senior member of the IEEE Computer Society and a member of the Mexican Computing Academy. His fields of interest include pattern recognition, digital image processing, augmented reality, and mechatronics.

Vianey Guadalupe Cruz Sánchez was born in Cárdenas, Tabasco, México, on September 14, 1978. She earned a B.Sc. degree in Computer Engineering from the Instituto Tecnológico de Cerro Azul, México, in 2000; the M.Sc. degree in computer science at the Center of Research and Technological Development (CENIDET) in 2004; and the Ph.D. in Computer Science from CENIDET in 2010. She is a Professor at the Autonomous University of Ciudad Juarez, Chihuahua, México. Prof. Cruz is a level-one member of the Mexican National Research System. She is a member of the IEEE Computer Society. Her fields of interest include neurosymbolic hybrid systems, digital image processing, knowledge representation, artificial neural networks, and augmented reality.

1 Introduction to Augmented Reality

1.1 CONTEXT

More than 30 years have passed since, in 1992, Thomas Caudell and David Mizell published the paper where the term "augmented reality" (AR) appeared for the first time (Caudell & Mizell, 1992). Thenceforth, the AR field has experimented with many changes. However, AR has not yet reached its consolidation. At the time of writing this book (2023), technology has advanced by significant steps. Furthermore, the pandemic accelerated digital transformation, and as a consequence, AR is becoming a common field for most people (Saleem et al., 2021).

At first, people believed that AR was purely fiction because it began to appear in movies like "Robocop," "Top Gun," "Ready Player One," and "Iron Man." However, with technological development, AR became a reality. It is common today to watch AR applied in sports (da Silva et al., 2021). For example, in car racing broadcasts, we can see the scene enriched with information regarding car speed and the physical distance to an opponent. In baseball, we can observe the strike area, or the yellow first and ten mark in football. In soccer, we can see a virtual line to verify the possibility of an offside. However, there is still some mistrust about the use, advantages, and real applications of AR.

Until a few years ago, adding virtual elements and integrating them into the real scene to transform the reality observed by a human seemed almost impossible. Therefore, Milgram et al. (1994), presented the Reality–Virtuality (RV) continuum as an attempt to describe a scale ranging from an entirely virtual environment to an entirely real environment. Figure 1.1 shows a diagram of the RV continuum. As observed, an environment comprising only virtual objects is defined on the right. Hence, all the objects are generated using a computer, and the object interaction happens using a computer peripheral. On the left, an environment containing only real objects is defined. Consequently, the environment includes real objects that can be touched and felt as it is commonly conducted in daily life.

From Figure 1.1, it is essential to highlight that a mixed-reality environment in which the real world and virtual objects interact is defined in the middle. On the AR side, virtuality augments reality; on the AV side, reality augments virtuality. Today, millions of people have access to AR experiences. Therefore, it is appropriate to define this fascinating technology.

DOI: 10.1201/9781003435198-1

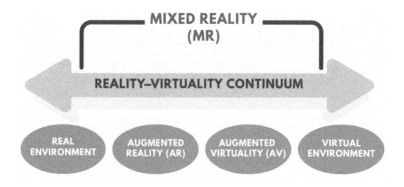

FIGURE 1.1   The scheme for the reality–virtuality (RV) continuum (Milgram et al., 1994).

1.2   AUGMENTED REALITY

In the literature, many definitions can be encountered regarding AR. However, The most critical fact that must be considered for AR is the action of combining real and virtual worlds in real time. Moreover, the mixture of real and virtual must be as transparent as possible to the user. According to Azuma (1997), an AR system must include at least three characteristics: (i) interactive in real time, (ii) registered in 3D, and (iii) combines real and virtual content. Therefore, AR is a technology that superimposes (includes) virtual objects in real scenarios that are observed through the screen of a technological device (Vergara et al., 2016). In order to design an AR system, at least four elements are needed:

1. A camera that can acquire digital images of the real world.
2. A device in which the images of the world can be projected together with the real and the virtual models, which can be a computer screen or a head-mounted display (HMD).
3. An element that allows interpreting the information obtained from the real world, recognizing the objects of interest, estimating the pose to superimpose the virtual objects, and performing the tracking algorithm, which is typically a central processing unit (CPU).
4. An object that allows the AR to be activated. In most cases, AR is detonated by a digital marker. The marker serves as a cue of where the virtual object will be inserted in the scene observed on the device screen.

Hence, the steps followed to experiment with AR are: (i) the user executes the AR application, (ii) the digital camera for capturing video is turned on, (iii) the captured video is displayed on the device screen, (iv) a valid marker is shown on the front of the camera, (v) the marker is recognized, (vi) the 2D/3D digital model corresponding to the marker detected is retrieved, (vii) pose estimation is calculated, (viii) the digital model is restructured according to the pose (registration), (ix) AR is activated; hence, the virtual object is inserted into the scene, and finally, (x) the marker tracking is started.

FIGURE 1.2 An example of AR scenes.

An example of AR is shown in Figure 1.2. As observed, the scene is acquired using the camera of a mobile device. Then, the video captured is displayed on the screen. When the marker appears on the scene, the virtual object is inserted where the marker is located, and the tracking process starts. If the marker is moved (left, right, close, or far) or is rotated, the tracking process computes the new position, rotation, and scale in which the virtual object is displayed.

AR has gone from being a figment of the imagination or science fiction to becoming a valuable product for various fields of science, especially for education. According to the object recognition technique, AR can be broadly divided into two main branches: marker-based AR and markerless AR (Oufqir et al., 2020). Marker-based AR employs a marker to activate the AR experience, while markerless AR does not need a fiducial marker.

1.2.1 MARKER-BASED AR

A traditional AR marker, or fiducial marker, is an object that must be detected, recognized, and tracked inside the AR scene. The marker is an image with a white background that can include different objects, shapes, or textual information. The marker is usually printed for easy handling. Then, the objects in the image are recognized using computer vision techniques that learn the features inside the marker, such as corners and edges (Boonbrahm et al., 2020).

The marker's location inside the image in x and y spatial coordinates determines where the virtual objects will be inserted. Therefore, the marker must contain features that can be easily detected, recognized, and tracked. Initially, the markers were designed as binary (black and white) images to reduce the recognition complexity. Later, with the advent of robust computer vision techniques, the markers can include almost any object, shape, and color.

For an AR system, a database of markers is constructed to contain all the valid markers to be recognized. Then, one virtual model is associated with each marker to build a dictionary. The database can be stored locally or in a cloud. Different sites, such as Brosvision s.r.o. (2023), offer support to create efficient markers. The designer must ensure that the markers used in an AR application contain sufficient features to

FIGURE 1.3   Examples of markers.

distinguish one from another. In addition, it must be ensured that the information in the markers is not so dense that it can be recognized even at far distances.

In Figure 1.3, four examples of markers are depicted. The first example shows a marker contained in a drop shape. The second shows a marker containing many colored lines, triangles, and quadrilaterals. The third example shows a quick response (QR) code. The last one is a traditional fiducial marker.

Currently, markers can appear on circular objects such as a bottle. Consequently, the number of applications in which AR can be used increases due to not having only flat markers.

1.2.2   MARKERLESS AR

Markerless AR does not need a marker to activate the AR experience. Instead, the system scans the environment to recognize a particular element inside the scene to determine where to overlay virtual models (Sadeghi & Soo, 2020). A flat surface, such as an intersection point, a table, a wall, or the floor, is typically employed to detonate the AR. Then, the virtual model is anchored on the surface to be observed at different sizes and perspectives. In summary, markerless AR eliminates the need for a tracking stage.

FIGURE 1.4 Example of markerless AR.

Prior knowledge of the environment where the virtual objects will be superimposed is not needed in markerless AR. Therefore, a critical stage in AR reality is the adequate alignment of the virtual object inside the real scene. This process is called registration. Registration implies the transformation of many data sets into a coordinate system. Particularly, markerless AR needs robust registration methods. One famous registration method is "simultaneous localization and mapping (SLAM)." SLAM can obtain real-time information from an unknown environment. Using the information obtained, SLAM can determine points that can be used to position virtual objects in the physical space. Because the technological devices employed to experience AR frequently include visual and inertial sensors, a variant of SLAM called Visual SLAM is employed (Jinyu et al., 2019).

Figure 1.4 depicts an example of a markerless AR system. As can be observed, no fiducial marker was employed to insert the virtual model of a taxi. Instead, the floor plane was employed as a cue to insert the virtual model.

In markerless AR, the digital content is overlaid, considering the geometry of an object. This kind of AR is popular for gaming because the user's range of motion increases. One of the most important examples is Pokémon GO; scientists considered the game an excellent resource for promoting physical activity. However, the massive use of AR games also helped to understand the main disadvantages of using AR, such as the limited field of view, attention tunneling, power consumption due to heavy processing, lack of flexibility, and the issue of addiction (Wang, 2021).

1.3 MOBILE AUGMENTED REALITY

Mobile computing can significantly complement AR technology. Therefore, developers have preferred implementing AR in mobile devices instead of personal computers (PCs) in recent years. The mixture of AR with mobile computing is called mobile AR (MAR). Gutiérrez et al. (2016) and Chen (2019) offered the following definition for MAR: "a real-time direct or indirect view of a real-world environment that has been augmented by adding virtual computer-generated information to it."

With MAR, several advantages are gained in comparison to PC-based AR, such as portability, use of the device camera for capturing the real-world view, powerful processors to recognize and track the objects of interest, the capability of rendering and displaying 3D graphics and video, and the use of the device sensors such as global positioning systems (GPS), inertial measurement units (IMU), accelerometers, and gyroscopes (Miranda et al., 2016).

In MAR, the mobile device can be an ultra-portable computer, a tablet, a smart telephone, or a portable videogame console. An example of the MAR experience is shown in Figure 1.5. The user observes the real scenario employing a mobile device's camera. Then, a cue, such as a building, is detected using computer vision techniques. Finally, the virtual object is inserted inside the scene.

MAR proposes a new form of interaction between systems and users. The user points the device's camera at a specific object or marker. Then, on the screen, the scene is represented with enriched information about the captured scenario. Besides portability, one of the main advantages of MAR is that it could provide a natural way to promote collaborative work between users. The collaborative feature of MAR allows multiple devices to intuitively discover and connect to share information and interact with the scene's elements to create a more natural and compelling environment that is especially useful for teaching and training purposes (Billinghurst & Kato, 2002). In collaborative MAR, multiple users must share at least one common place inside the augmented environment. According to Lukosch et al. (2015), users can collaborate face-to-face, remotely, or by combining both. Each user has a view of the private or shared objects in the augmented application.

One of the main challenges to solve in MAR is the effective use of computer vision techniques since most of the processes must be performed actively. Active vision implies that a camera's point of view is manipulated to observe the environment and obtain richer information, as the human biological vision system naturally performs it. In addition, challenges regarding power consumption, memory limitations, delays, and display requirements should be surpassed.

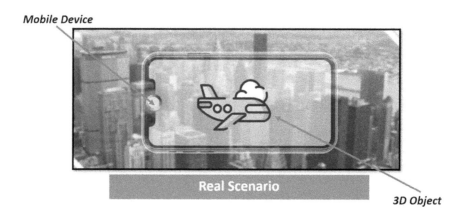

FIGURE 1.5   An example of mobile augmented reality.

1.3.1 GEOLOCATION-BASED AUGMENTED REALITY

When the AR experience succeeds in concrete exterior places, it is called geolocation-based markerless AR. This kind of AR succeeds by using the information of the physical space in which the user is located. The virtual model is inserted into the scene

FIGURE 1.6   Geolocation-based AR.

utilizing the information recovered from the mobile device's GPS, compass, and accelerometer.

According to Erra and Capece (2019), geolocation-based AR retrieves the mobile device's physical location and computes the point of interest, and then a virtual model is overlaid into the real scene viewed in the device display.

Geolocation-based AR is beneficial for travelers inside a specific area because the virtual information inserted into the scene can help them understand the environment in which they are located.

Figure 1.6 shows an example of a geolocation-based AR. The user points the mobile device camera to the surrounding environment to start the experience. Then, a location is retrieved using the device sensors, obtaining latitude and longitude information. As a result, the application displays information about the point of interest and shows a picture of the place where the user is located.

One of the main problems with geolocation-based AR is the GPS error. Currently, there is an error of about 7–13 meters in urban environments (Merry & Bettinger, 2019). The Wikitude app, launched in 2008, was the first commercial application that employed the AR geolocation-based approach.

1.4 THE HISTORY OF AUGMENTED REALITY

Although AR is considered a new technology, its origin dates back to 1901 when Baum (1901) presented "Master Key," a novel that describes the adventures of a boy who experiments with electricity. In the story, the boy finds the key to electricity and meets the demon that guards it. The demon gives him a gift called the "character maker," which allows him to judge people. The device was considered the prototype of what today is called AR and consisted of glasses that overlaid information regarding the people the boy meets. People's foreheads were marked with a letter indicating their character, including G for good, E for evil, W for wise, and C for cruel.

Years later, in 1962, the cinematographer Heilig (1962) presented "Sensorama," considered the first immersion machine. Sensorama offers users a multisensorial immersive cinematographic experience regarding driving a motorcycle. The machine included images, sound, seat vibration, the feeling of wind on the face, and the odor of the environment in which the motorcycle was driven. In 1964, Sutherland presented "Sketchpad" as part of his Ph.D. dissertation. The system allows a person and a computer to converse using line drawings. Sketchpad employed a TX-2 computer and a light pen to manipulate graphics objects directly. Users can directly draw points, line segments, circular arcs, and duplicate drawings on the computer monitor (Sutherland, 1964). In 1968, Sutherland presented the head-mounted three-dimensional display called "Sword of Damocles." The device allows the user to present a perspective image that changes as he moves. By placing 2D images on the observer's retina, an illusion of seeing a 3D object was generated. The head-mounted display (HMD) was a machine with a suspending counterbalance mechanical arm and ultrasonic transducers to track the head movement (Sutherland, 1968). In 1969, Furness (the grandfather of virtual reality) started developing the "Super-Cockpit" for the US Air Force. The cockpit was intended to aid the piloting of combat aircraft by projecting computer-generated three-dimensional maps, radar and infrared images, and aeronautical data in real time (Furness, 1969). The work continued until 1989, when Furness left the US Air Force.

In 1975, Krueger developed the artificial intelligence laboratory called "Videoplace," which is considered the first VR interface. Videoplace offers users live video images with a computer graphic world for real-time manipulation and interaction of virtual objects. Video cameras and projectors were employed to offer the user interaction with silhouettes (Krueger et al., 1985; Krueger, 1977). The Massachusetts Institute for Technology (MIT) presented the "Aspen Movie Map" in 1978. In the virtual simulation, the users can navigate the streets of Aspen, Colorado, in three different modes: summer, winter, and polygons. The summer and winter models were based on real pictures, while the polygon model included a basic 3D city model (Lippman, 1980). In search of creating an HMD that reduces the price and size of a military flight simulator, McDonnell-Douglas Corporation presented "The Vital Helmet" in 1979. The helmet was considered the first example of a VR HMD. The helmet used a head tracker to determine where the pilot was observing and cathode ray tubes to project images onto a beam splitter in front of the pilot's eyes. The pilot can manipulate the cockpit controls while seeing computer-generated images of the real world (Macleod & Coblintz, 1979).

In 1980, Steve Mann, known as "the Father of Wearable Computing," presented the first wearable AR device called the "EyeTap." The monocular device combines a head-up display and a camera to show computer-generated images inside the scenes registered by the user. The images alter the environment's visual perception, creating AR (Mann, 2013). In 1981, programmer Dave Raggett presented the Virtual Reality Modeling Language (VRML) as a proposal to expand the World Wide Web (WWW) to support virtual reality regardless of the platform used. VRML allows the representation of interactive 3D graphics in a web browser (Raggett, 2015). In 1982, Dan Reitan and his team made the first meteorological broadcasting as a precursor concept of AR. The idea was to take geospatial maps to overlay multiple weather radar images (Uma, 2019). The application remains in wide use today. The National Aeronautics and Space Agency (NASA) created a liquid crystal display (LCD)-based HMD called the "Virtual Visual Environment Display (VIVED)" in 1985. This device comprises the LCDs of a Sony Watchman, a DEC PDP 11-40 computer, a graphic computer, and a noncontact tracker. The tracker measured the movement of the user's head, and the data were transmitted to the DEC PDP 11-40. Then, the information was passed to the graphics computer to calculate the images to be displayed (Billinghurst et al., 2015). Also, in 1985, Jaron Lanier, Scott Fisher, and Thomas Zimmerman founded VPL Research, Inc., the first company that sold virtual reality glasses and gloves (Conn et al., 1989). In 1987, Ron Feigenblatt from IBM presented an application that uses a small flat panel display in which objects can be oriented and positioned by hand (Feigenblatt, 1987). The application laid the foundations for manipulating objects on smart devices today. VIVED was the base for a new NASA project called "Virtual Interface Environment Workstation" (VIEW). The VIEW was presented in 1988 and consisted of a head-mounted stereoscopic display controlled by voice and gestures. The users can observe the 360-degree synthesized environment and interact with its components (Fisher et al., 1988). In 1989, Jaron Lanier coined the term "virtual reality" (Lanier, 1992). In the same year, Furness left the Air Force to founding the Human Interface Technology (HIT) Laboratory at the University of Washington. The HIT laboratory is considered one of the most influential research groups regarding VR and AR (Billinghurst et al., 2015).

Caudell and Mizell presented in 1990 a project for helping Boeing engineers connect cables on an airplane control panel. The system shows which wires should be connected using AR. Unfortunately, the project was not well received. Nevertheless, as a result, the first academic paper containing the term "augmented reality" was published (Caudell & Mizell, 1992). In 1992, Louis Rosenberg developed "Virtual Fixtures," the first fully immersive AR system. This system overlaid information in a work area aiming to improve human productivity. Virtual Fixtures allowed military personnel to control and guide machinery virtually. Also, the system is an example of telerobotics because, using an exoskeleton, the user controls the robot's movements (Rosenberg, 1993). In 1993, Feiner and a team at Columbia University presented Knowledge-based Augmented Reality for Maintenance Assistance (KARMA). In KARMA, a see-through head-mounted display is employed to explain maintaining a laser printer (Feiner et al., 1993). The Reality–Virtuality (RV) continuum was presented in 1994 by Milgram et al. (1994). Also, 1994 was the year when the first AR theater production called "Dancing in Cyberspace" was presented. In the production, acrobats danced around virtual objects projected on stage (Aggarwal & Singhal, 2019). In 1996, the "Glow Puck" converted ice hockey into a more watchable sport. Glow Puck allows viewers to follow the puck easily. A blue trail is projected when the puck travels slowly, and a red trail is projected when the puck travels fast (Chad, 2021). Feiner et al. (1997) presented the "Touring Machine," the first MAR system. The machine serves as a campus information system, helping find places and perform queries regarding buildings. The "Touring Machine" comprises a head-tracked, see-through, head-worn, 3D display, and a 2D display with a stylus. Also, in 1997, a survey of AR that introduced a definition of AR and summarized most work conducted was presented (Azuma, 1997). The survey became one of the benchmarks in the field of AR. Currently, it has more than 15,000 citations. The first edition of the IEEE/ACM International Workshop on Augmented Reality (IWAR) was conducted in 1998 (Reinhold et al., 1999). IWAR was the first academic conference that specialized in the field of AR. In September 1998, Sportvision showed for the first time a new AR technology, the "1st and Ten Line." The technology projected a yellow first-down line visible only to TV viewers, which today is a crucial part of the game (Chad, 2021). In 1999, the University of Washington presented an open-source C/C++ library for making AR applications called "Artoolkit." The library includes functions for video capturing and employing computer vision techniques to find and track the square markers (Kato, 2002). Today, Artoolkit is one of the most widely used AR libraries. The first edition of the International Symposium on Mixed Reality (ISMR) was conducted in 1999 in Yokohama, Japan (Billinghurst & Kato, 1999).

After two successful IWAR editions, it became a symposium, resulting in the International Symposium on Augmented Reality (ISAR) in 2000 (Klinker et al., 2000). The same year, Bruce Thomas from the University of South Australia developed the first-person outdoor/indoor application called "AR Quake." The architecture comprises an accurate six degrees of freedom tracking system based on GPS, digital compass, and fiducial vision-based tracking (Thomas et al., 2002). Azuma et al. (2001) published a survey to complement the work presented in 1997 (Azuma, 1997). Currently, it has more than 5000 citations. The ISMR and the ISAR conferences were combined into the IEEE and ACM International Symposium on Mixed and Augmented Reality (ISMAR) in 2002 (Stricker et al., 2002). Nowadays, ISMAR is the most important conference in the AR field. Wagner and Schmalstieg (2003) presented the first AR system with self-tracking running on a personal digital assistant (PDA) with a commercial camera. The system employs ARtoolkit to offer the users a 3D augmented view of the environment. The tracking task was conducted on a client/server architecture to provide better performance. In 2004, the first system for tracking 3D markers on mobile phones was presented by Mohring et al. (2004). This development allowed for the detection and differentiation of different 3D markers and the integration of 3D renderings into a live video stream using an OpenGL rendering pipeline. In the same year, the " Human Pacman" AR game was presented (Cheok et al., 2004). This game employs GPS and an accelerometer to determine where the Pacman and the ghosts will be located in a wide outdoor physical area. In 2006, Nokia initiated the Mobile Augmented Reality Applications (MARA) project (Schmeil & Broll, 2006). MARA utilizes an accelerometer, a GPS, and a compass to overlay information to the images observed on the screen. In 2007, Sony presented the AR game "The Eye of the Judgement." This game uses AR to mix physical cards and virtual creatures that come to life when using the PlayStation camera (Billinghurst et al., 2015). The same year, Apple presented the first iPhone (Goggin, 2009). In 2008, the first commercial Android device called "HTC Dream" was presented (Butler, 2011). In the same year, Mobilizy launched Wikitude, an AR browser for Android devices that combines data from GPS and a compass with registers from Wikipedia that substitute virtual maps with real images of a row (Lo & Gong, 2020). SPRXmobile launched in 2009 a similar application to Wikitude called "Layar," which is employed to create and share AR digital content. Layar can display points of interest and user annotations based on the mobile device sensors (Liao & Humphreys, 2015). The Esquire magazine provided AR content on its cover and 12 pages in the same year. Using costume software and pointing a webcam at the magazine cover, readers can observe a video of Robert Downey-Jr (Ikonen & Uskali, 2020).

In 2010, a peripheral for Xbox 360 named "Kinect" was presented by Microsoft. The Kinect was an accessory that allowed one to interact with a console without having physical contact (Bostanci et al., 2015). The appearance of the Kinect triggered the development of many AR applications. In 2011, "Aurasma Lite" was launched as an AR application for iPhone and Android. Aurasma made it easy to create AR content. Defining and associating a marker with a 3D model was necessary to create an attractive educational experience (Rocha & Nereu, 2017). Later, Aurasma was renamed "HP Reveal" but is currently discontinued. In the same year, "Vuforia" was released. Vuforia recognizes and tracks planar images and 3D objects in real time. Vuforia is one of the most used AR platforms worldwide (Sendari et al., 2020). Google presented in 2012 "Google Glass." This wearable technology included a small screen to display various information such as emails, text messages, and weather forecasts. Most AR applications employing Google Glass serve for navigation (Rehman & Cao, 2017). In 2013, Volkswagen launched the Mobile Augmented Reality Technical Assistance (MARTA) project to facilitate their customers in inspection and maintenance tasks. MARTA aimed to substitute the classical automobile paper manual. MARTA uses a mobile device's camera to view and label the internal parts of a vehicle (Chouchene et al., 2022). An AR Android platform named "Tango" was launched by Google in 2014. Tango employed computer vision and image-processing techniques to visualize how the furniture would look in a room, draw surfaces in a house, or personalize videos with 3D filters. Motion tracking, area learning, and depth perception are the core technologies Tango employs (Marder, 2016). In the same year, the authors of this book opened the "Computer Vision and Augmented Reality Laboratory" at Universidad Autonoma de Ciudad Juarez. In 2015, Microsoft announced the AR headset called "Hololenses." Hololenses allow adding computer images (holograms) to the physical environment observed through a digital display (Evans et al., 2017; Xue et al., 2019). The emergence of Pokémon GO marked a before and after in AR history. The game was created by Niantic company in 2016. Users must navigate in an environment to find labels where Pokémon characters are hidden. Once a Pokémon is detected, the user must try to catch it. Also, the users should sustain battles with other characters in places called gyms (Wang, 2021). In 2017, Apple released the AR development framework "ARKit" for iOS devices. With ARKit, the user does not have to do any calibration. Moreover, ARKit allows for creating a map of the area, detecting tabletops and floors, and locating the mobile device in a physical space (Wang, 2018). In the same year, Facebook (now Meta) released "Spark AR Studio" to create virtual and augmented reality effects that can be applied to faces (Javornik et al., 2022). In 2018, the Tango project ended to give a step to "ARCore." ARCore allows the creation of AR content for Android devices. ARCore is supported by three fundamental pillars: tracking, environmental understanding, and light estimation (Lu et al., 2021). In 2019, the HoloLens 2 was presented. Compared with HoloLens 1, the HoloLens 2 offers a wide field of view, spatial mapping, natural interaction, spatial sound, voice recognition, computer vision, and biometric authentication (Tu et al., 2021).

In 2020, Nintendo released the location-based AR game "Mario Kart Live: Home Circuit." Mario Kart employs a racing toy car controlled by a remote joystick, and a streaming onboard video is obtained using a camera. The virtual racing playsets are built around the house, and the traditional objects are virtually included (Kerdvibulvech, 2021). In 2021, Facebook changed its name to "Meta," and in 2022, Spark AR Studio changed its name to "Meta Spark Studio" (Afshar, 2023). In 2023, Apple announced the "Vision Pro," a headset that blends digital content with the physical world. Also, the VisionOs was launched, which is the first spatial operating system (Waisberg et al., in press).

We are not even in the middle of the 2020s at the time of writing this book. However, it is expected that AR will shortly achieve its consolidation and become a daily use tool for all those with a technological device.

1.5 AUGMENTED REALITY VS. VIRTUAL REALITY

AR and VR terms are frequently confused even when several characteristics differentiate them. Unlike VR, AR does not simulate reality. Instead, AR superimposes contextual data without altering reality. As a result, AR systems enable the user to examine the physical world while additional object information is presented on the screen of a technological device (Barraza et al., 2015).

On the other hand, VR employs a human–computer interface to simulate a realistic environment. The virtual scenes and objects created using computers give the user the feeling of being immersed in the virtual world (Zheng et al., 1998). A wearable device must be used if a user wants to experiment with VR. By observing the Reality–Virtuality (RV) continuum, AR is close to real environments, while VR is located on the other extreme, close to the virtual environments.

In summary, complete immersion in an entirely artificial environment is experimented with in VR. The users must wear a headset that completely covers their eyes. On the other hand, the real world is enhanced with virtual objects in AR. AR does not need complex equipment and can be experimented with using a smartphone.

Figure 1.7 shows an example of AR and VR. On the left, a marker-based application is shown. As observed, a fiducial marker was located on a laptop keyboard. Computer vision techniques recognize the marker, and the robot 3D model is inserted into the scene. The keyboard and the marker are real, while the robot is virtual. The device's camera should be moved if the user wants to observe the robot from a different perspective. Otherwise, the marker should be rotated, zoomed in or out, or translated. On the right, a VR scene is shown. All the objects, the wall, the floor, and the robot are virtual. The objects can be manipulated and observed from different angles and scales using head tracking information or a peripheral such as a mouse or a joystick.

In VR, the 3D world to look around can be entirely imaginary or a replica of the real world, and the experience frequently includes visual and auditory cues. In AR, 2D or 3D virtual objects are overlaid in the real world. The immersion is not

FIGURE 1.7   Augmented reality vs. virtual reality.

TABLE 1.1
Summary of the Main Features of Augmented Reality and Virtual Reality

Augmented Reality	Virtual Reality
The system augments the real-world scene (partial immersion)	The virtual environment is fully immersive
The user maintains a sense of presence in the real world	The senses are under the control of the system
It needs a mechanism for recording the real world with virtual objects	It needs a mechanism to show the virtual world to the user
The process for recording the real and virtual is complex	It is challenging to create a virtual world
Less computing power	More computing power.
Augments the reality	Replaces the reality
Minimal rendering requirements	Requires very realistic images
High accuracy tracking	Low accuracy tracking
Require less complex technological devices	Requires complex technological devices

total because the user always observes the real world (Jung & Tom, 2017). Table 1.1 summarizes the main features of AR and VR. In summary, AR and VR are two powerful technological tools with differences and similarities commonly adopted in industry and education.

REFERENCES

Afshar, J. (2023). *Hands-On Augmented Reality Development with Meta Spark Studio: A Beginner's Guide* (1st ed.). Apress.

Aggarwal, R., & Singhal, A. (2019). Augmented Reality and its Effect on Our Life. *Proceedings of the 9th International Conference on Cloud Computing, Data Science & Engineering (Confluence)*, 510–515. https://doi.org/10.1109/CONFLUENCE.2019.8776989

Azuma, R. (1997). A Survey of Augmented Reality. *Presence: Teleoperators and Virtual Environments*, 6(4), 355–385. https://doi.org/10.1162/pres.1997.6.4.355

Azuma, R., Baillot, Y., Behringer, R., Feiner, S., Julier, S., & MacIntyre, B. (2001). Recent Advances in Augmented Reality. *IEEE Computer Graphics and Applications*, 21(6), 34–47. https://doi.org/10.1109/38.963459

Barraza, R., Vergara, O., & Cruz, V. (2015). A Mobile Augmented Reality Framework Based on Reusable Components. *IEEE Latin America Transactions*, 13(3), 713–720. https://doi.org/10.1109/TLA.2015.7069096

Baum, L. (1901). *The Master Key: An Electrical Fairy Tale Founded Upon the Mysteries of Electricity and the Optimism of its Devotees* (1st ed.). Bowen-Merrill Company.

Billinghurst, M., Clark, A., & Lee, G. (2015). A Survey of Augmented Reality. *Foundations and Trends in Human–Computer Interaction*, 8(2–3), 73–272. https://doi.org/10.1561/1100000049

Billinghurst, M., & Kato, H. (1999). First International Symposium on Mixed Reality (ISMR). *Proceedings of the First International Symposium on Mixed Reality (ISMR)*, 261–264.

Billinghurst, M., & Kato, H. (2002). Collaborative Augmented Reality. *Communications of the ACM, 45*(7), 64–70. https://doi.org/10.1145/514236.514265

Boonbrahm, S., Boonbrahm, P., & Kaewrat, C. (2020). The Use of Marker-Based Augmented Reality in Space Measurement. *Procedia Manufacturing, 42,* 337–343. https://doi.org/10.1016/j.promfg.2020.02.081

Bostanci, E., Kanwal, N., & Clark, A. F. (2015). Augmented Reality Applications for Cultural Heritage Using Kinect. *Human-Centric Computing and Information Sciences, 5*(1), 1–18. https://doi.org/10.1186/s13673-015-0040-3

Brosvision s.r.o. (2023, February). *Augmented Reality Marker Generator – Brosvision.* www.brosvision.com/ar-marker-generator/

Butler, M. (2011). Android: Changing the Mobile Landscape. *IEEE Pervasive Computing, 10*(1), 4–7. https://doi.org/10.1109/MPRV.2011.1

Caudell, T., & Mizell, D. (1992). Augmented Reality: An Application of Heads-Up Display Technology to Manual Manufacturing Processes. *Proceedings of the Twenty-Fifth Hawaii International Conference on System Sciences (HICSS),* 659–698. https://doi.org/10.1109/HICSS.1992.183317

Chad, G. (2021). *Augmented Reality in Sport Broadcasting* [Ph.D.]. Virginia Commonwealth University.

Chen, Y. (2019). Effect of Mobile Augmented Reality on Learning Performance, Motivation, and Math Anxiety in a Math Course. *Journal of Educational Computing Research, 57*(7), 1695–1722. https://doi.org/10.1177/0735633119854036

Cheok, A., Goh, K., Liu, W., Farbiz, F., Teo, S., Teo, H., Lee, S., Li, Y., Fong, S., & Yang, X. (2004). Human Pacman: A Mobile Wide-Area Entertainment System Based on Physical, Social, and Ubiquitous Computing. *Proceedings of the ACM SIGCHI International Conference on Advances in Computer Entertainment Technology,* 360–361. https://doi.org/10.1145/1067343.1067402

Chouchene, A., Ventura, A., Charrua, F., & Barhoumi, W. (2022). Augmented Reality-Based Framework Supporting Visual Inspection for Automotive Industry. *Applied System Innovation, 5*(3), 1–13. https://doi.org/10.3390/asi5030048

Conn, C., Lanier, J., Minsky, M., Fisher, S., & Druin, A. (1989). Virtual Environments and Interactivity: Windows to the Future. *ACM SIGGRAPH Computer Graphics, 23*(5), 7–18. https://doi.org/10.1145/77276.77278

da Silva, A., Albuquerque, G., & de Medeiros, F. (2021). A Review on Augmented Reality Applied to Sports. *Proceedings of the 16th Iberian Conference on Information Systems and Technologies (CISTI),* 1–6. https://doi.org/10.23919/CISTI52073.2021.9476570

Erra, U., & Capece, N. (2019). Engineering an Advanced Geo-Location Augmented Reality Framework for Smart Mobile Devices. *Journal of Ambient Intelligence and Humanized Computing, 10*(1), 255–265. https://doi.org/10.1007/s12652-017-0654-6

Evans, G., Miller, J., Pena, M., MacAllister, A., & Winer, E. (2017). Evaluating the Microsoft HoloLens Through an Augmented Reality Assembly Application. *Proceedings of the Degraded Environments: Sensing, Processing, and Display,* 1–16. https://doi.org/10.1117/12.2262626

Feigenblatt, R. (1987). Absolute Display Window Mouse/Mice. *IBM Technical Disclosure Bulletin, 29*(10), 1–15.

Feiner, S., MacIntyre, B., Höllerer, T., & Webster, A. (1997). A Touring Machine: Prototyping 3D Mobile Augmented Reality Systems for Exploring the Urban Environment. *Personal Technologies, 1*(4), 208–217. https://doi.org/10.1007/BF01682023

Feiner, S., Macintyre, B., & Seligmann, D. (1993). Knowledge-Based Augmented Reality. *Communications of the ACM, 36*(7), 53–62. https://doi.org/10.1145/159544.159587

Fisher, S., Wenzel, E., Coler, C., & McGreevy, M. (1988). Virtual Interface Environment Workstations. *Proceedings of the Human Factors Society Annual Meeting*, *32*(2), 91–95. https://doi.org/10.1177/154193128803200219

Furness, L. (1969). *The Application of Head-Mounted Displays to Airborne Reconnaissance and Weapon Delivery*. Wright-Patterson Air Force Base, Ohio, USA.

Goggin, G. (2009). Adapting the Mobile Phone: The iPhone and its Consumption. *Journal of Media & Cultural Studies*, *23*(2), 231–244. https://doi.org/10.1080/1030431080 2710546

Gutiérrez, E., Jiménez, F., Ariza, A., & Taguas, J. (2016). DiedricAR: A Mobile Augmented Reality System Designed for the Ubiquitous Descriptive Geometry Learning. *Multimedia Tools and Applications*, *75*(16), 9641–9663. https://doi.org/10.1007/s11042-016-3384-4

Heilig, M. (1962). *Sensorama Simulator* (Patent US3050870). https://patents.google.com/pat ent/US3050870A/en

Ikonen, P., & Uskali, T. (2020). Augmented Reality as News. In Turo Uskali, Astrid Gynnild, Sarah Jones, Esa Sirkkunen (eds) *Immersive Journalism as Storytelling* (pp. 147–160). Routledge. https://doi.org/10.4324/9780429437748-16

Javornik, A., Marder, B., Barhorst, J., McLean, G., Rogers, Y., Marshall, P., & Warlop, L. (2022). 'What Lies Behind the Filter?' Uncovering the Motivations for Using Augmented Reality (AR) Face Filters on Social Media and their Effect on Well-Being. *Computers in Human Behavior*, *128*, 1–12. https://doi.org/https://doi.org/ 10.1016/j.chb.2021.107126

Jinyu, L., Bangbang, Y., Danpeng, C., Nan, W., Guofeng, Z., & Hujun, B. (2019). Survey and Evaluation of Monocular Visual-Inertial SLAM Algorithms for Augmented Reality. *Virtual Reality & Intelligent Hardware*, *1*(4), 386–410. https://doi.org/https://doi.org/ 10.1016/j.vrih.2019.07.002

Jung, T., & Tom, M. (2017). *Augmented Reality and Virtual Reality: Empowering Human, Place and Business* (1st ed.). Springer Publishing Company.

Kato, H. (2002). ARToolKit: Library for Vision-based Augmented Reality. *Institute of Electronics, Information and Communication Engineers (IEICE) Technical Report*, *101*(652), 79–86.

Kerdvibulvech, C. (2021). Location-Based Augmented Reality Games Through Immersive Experiences. In Schmorrow, D., & Fidopiastis, D. (Eds.), *Augmented Cognition* (pp. 452–461). Springer International Publishing.

Klinker, G., Behringer, R., Mizell, D., & Navab, N. (2000). IEEE and ACM International Symposium on Augmented Reality (ISAR). *Proceedings of the IEEE and ACM International Symposium on Augmented Reality (ISAR)*, 1–200. https://doi.org/10.1109/ ISAR.2000.880915

Krueger, M. (1977). Responsive Environments. *Proceedings of the National Computer Conference*, 423–433. https://doi.org/10.1145/1499402.1499476

Krueger, M., Gionfriddo, T., & Hinrichsen, K. (1985). VIDEOPLACE – An Artificial Reality. *Proceedings of the SIGCHI Conference on Human Factors in Computing Systems*, 35–40. https://doi.org/10.1145/317456.317463

Lanier, J. (1992). Virtual Reality: The Promise of the Future. *Interactive Learning International*, *8*(4), 275–279.

Liao, T., & Humphreys, L. (2015). Layar-ed Places: Using Mobile Augmented Reality to Tactically Reengage, Reproduce, and Reappropriate Public Space. *New Media & Society*, *17*(9), 1418–1435. https://doi.org/10.1177/1461444814527734

Lippman, A. (1980). Movie-Maps: An Application of the Optical Videodisc to Computer Graphics. *Proceedings of the 7th Annual Conference on Computer Graphics and Interactive Techniques*, 32–42. https://doi.org/10.1145/800250.807465

Lo, J., & Gong, G. (2020). Touring System using Augmented Reality – A Case Study of Yilan Cultural Industries. *Proceedings of the 3rd IEEE International Conference on Knowledge Innovation and Invention (ICKII)*, 204–207. https://doi.org/10.1109/ICK II50300.2020.9318893

Lu, F., Zhou, H., Guo, L., Chen, J., & Pei, L. (2021). An ARCore-Based Augmented Reality Campus Navigation System. *Applied Sciences*, *11*(16), 1–16. https://doi.org/10.3390/app11167515

Lukosch, S., Billinghurst, M., Alem, L., & Kiyokawa, K. (2015). Collaboration in Augmented Reality. *Computer Supported Cooperative Work: The Journal of Collaborative Computing (Online)*, *24*(6), 515–525. https://doi.org/10.1007/s10606-015-9239-0

Macleod, S., & Coblintz, D. (1979). *Visually Coupled System – Computer Generated Imagery (VCS-CGI) Engineering Interface*. Air Force Aerospace Medical Research Laboratory.

Mann, S. (2013). Vision 2.0. *IEEE Spectrum*, *50*(3), 42–47. https://doi.org/10.1109/MSPEC.2013.6471058

Marder, E. (2016). Project Tango. *ACM SIGGRAPH 2016 Real-Time Live!*, 1–25. https://doi.org/10.1145/2933540.2933550

Merry, K., & Bettinger, P. (2019). Smartphone GPS Accuracy Study in an Urban Environment. *PLOS One*, *14*(7), 1–19. https://doi.org/10.1371/journal.pone.0219890

Milgram, P., Takemura, H., Utsumi, A., & Kishino, F. (1994). Augmented Reality: A Class of Displays on the Reality-Virtuality Continuum. *Telemanipulator and Telepresence Technologies*, *2351*, 282–292. https://doi.org/10.1117/12.197321

Miranda, E., Vergara, O., Cruz, V., García, J., & Favela, J. (2016). Study on Mobile Augmented Reality Adoption for Mayo Language Learning. *Mobile Information Systems*, *2016*, 1–15. https://doi.org/10.1155/2016/1069581

Mohring, M., Lessig, C., & Bimber, O. (2004). Video See-Through AR on Consumer Cell-Phones. *Proceedings of the 3rd IEEE/ACM International Symposium on Mixed and Augmented Reality (ISMAR)*, 252–253. https://doi.org/10.1109/ISMAR.2004.63

Oufqir, Z., El Abderrahmani, A., & Satori, K. (2020). From Marker to Markerless in Augmented Reality. In Bhateja, V., Satapathy, S., and Satori, H. (Eds.), *Embedded Systems and Artificial Intelligence* (pp. 599–612). Springer Singapore.

Raggett, D. (2015). The Web of Things: Challenges and Opportunities. *Computer*, *48*(5), 26–32. https://doi.org/10.1109/MC.2015.149

Rehman, U., & Cao, S. (2017). Augmented-Reality-Based Indoor Navigation: A Comparative Analysis of Handheld Devices Versus Google Glass. *IEEE Transactions on Human-Machine Systems*, *47*(1), 140–151. https://doi.org/10.1109/THMS.2016.2620106

Reinhold, B., Gudrun, K., & David, M. (1999). International Workshops on Augmented Reality. *Proceedings of the International Workshops on Augmented Reality (1st ed)*. IEEE.

Rocha, V., & Nereu, T. (2017). Aurasma: A Tool for Education. Proceedings *of the 19th Symposium on Virtual and Augmented Reality (SVR)*, 257–260. https://doi.org/10.1109/SVR.2017.40

Rosenberg, L. (1993). Virtual Fixtures: Perceptual Tools for Telerobotic Manipulation. *Proceedings of IEEE Virtual Reality Annual International Symposium*, 76–82. https://doi.org/10.1109/VRAIS.1993.380795

Sadeghi, A., & Soo, C. (2020). A Survey of Marker-less Tracking and Registration Techniques for Health & Environmental Applications to Augmented Reality and Ubiquitous Geospatial Information Systems. *Sensors*, *20*(10), 1–26. https://doi.org/10.3390/s20102997

Saleem, M., Kamarudin, S., Shoaib, H., & Nasar, A. (2021). Influence of Augmented Reality App on Intention Towards E-learning Amidst COVID-19 Pandemic. *Interactive Learning Environments*, 1–15. https://doi.org/10.1080/10494820.2021.1919147

Schmeil, A., & Broll, W. (2006). MARA: An Augmented Personal Assistant and Companion. *ACM SIGGRAPH 2006 Sketches*, 141–es. https://doi.org/10.1145/1179849.1180025

Sendari, S., Firmansah, A., & Aripriharta. (2020). Performance Analysis of Augmented Reality Based on Vuforia Using 3D Marker Detection. *Proceedings of the 4th International Conference on Vocational Education and Training (ICOVET)*, 294–298. https://doi.org/10.1109/ICOVET50258.2020.9230276

Stricker, D., Müller, S., & Schmalstieg, D. (2002). First IEEE and ACM International Symposium on Mixed and Augmented Reality (ISMAR). *Proceedings of the First IEEE and ACM International Symposium on Mixed and Augmented Reality (ISMAR) (1ˢᵗ ed.)*. IEEE.

Sutherland, I. (1964). Sketch Pad a Man-Machine Graphical Communication System. *Proceedings of the SHARE Design Automation Workshop*, 6.329–6.346. https://doi.org/10.1145/800265.810742

Sutherland, I. (1968). A Head-Mounted Three Dimensional Display. *Proceedings of the December 9-11, 1968, Fall Joint Computer Conference, Part I*, 757–764. https://doi.org/10.1145/1476589.1476686

Thomas, B., Close, B., Donoghue, J., Squires, J., De Bondi, P., & Piekarski, W. (2002). First Person Indoor/Outdoor Augmented Reality Application: ARQuake. *Personal and Ubiquitous Computing*, *6*(1), 75–86. https://doi.org/10.1007/s007790200007

Tu, P., Gao, Y., Lungu, A., Li, D., Wang, H., & Chen, X. (2021). Augmented Reality Based Navigation for Distal Interlocking of Intramedullary Nails Utilizing Microsoft HoloLens 2. *Computers in Biology and Medicine*, *133*, 1–10. https://doi.org/https://doi.org/10.1016/j.compbiomed.2021.104402

Uma, S. (2019). Latest Research Trends and Challenges of Computational Intelligence Using Artificial Intelligence and Augmented Reality. In Anandakumar, H., Arulmurugan, R., & Onn, C. (Eds.), *Computational Intelligence and Sustainable Systems: Intelligence and Sustainable Computing* (pp. 43–59). Springer International Publishing. https://doi.org/10.1007/978-3-030-02674-5_3

Vergara, O., Cruz, V., Rodríguez, R., & Nandayapa, M. (2016). Recent ADvances in Augmented Reality (RADAR). *International Journal of Combinatorial Optimization Problems and Informatics*, *7*(3), 1–6.

Wagner, D., & Schmalstieg, D. (2003). First Steps Towards Handheld Augmented Reality. *Proceedings of the Seventh IEEE International Symposium on Wearable Computers, 2003*, 127–135. https://doi.org/10.1109/ISWC.2003.1241402

Waisberg, E., Ong, J., Masalkhi, M., Zaman, N., Sarker, P., Lee, A., & Tavakkoli, A. (in press). Apple Vision Pro and Why Extended Reality will Revolutionize the Future of Medicine. *Irish Journal of Medical Science (1971)*. https://doi.org/10.1007/s11845-023-03437-z

Wang, A. (2021). Systematic Literature Review on Health Effects of Playing Pokémon Go. *Entertainment Computing*, *38*, 1–10. https://doi.org/https://doi.org/10.1016/j.entcom.2021.100411

Wang, W. (2018). Understanding Augmented Reality and ARKit. In Wang, W. (ed.) *Beginning ARKit for iPhone and iPad: Augmented Reality App Development for iOS* (pp. 1–17). Apress. https://doi.org/10.1007/978-1-4842-4102-8_1

Xue, H., Sharma, P., & Wild, F. (2019). User Satisfaction in Augmented Reality-Based Training Using Microsoft HoloLens. *Computers*, *8*(1), 1–23. https://doi.org/10.3390/computers8010009

Zheng, J., Chan, K., & Gibson, I. (1998). Virtual Reality. *IEEE Potentials*, *17*(2), 20–23. https://doi.org/10.1109/45.666641

2 Augmented Reality Application Fields

2.1 CONTEXT

The main goal of AR is to "augment everything everywhere for everyone." As visualization technology, AR plays a crucial role in transitioning to a digital society. However, to this day, much research and work needs to be done to demonstrate the impact of AR applications on any industry or environment.

AR is recommended when a phenomenon cannot be simulated in reality for various reasons, such as the lack of a tangible object or when it is difficult to understand the object of study. Moreover, Azuma et al. (2001) explained the primary settings where AR could be implemented, including industry, education, medicine, maintenance, repair, annotation, military, robotics, and entertainment.

The following subsections show some of the efforts made by the authors of this book in the fields of education, medicine, and industry during the last 10 years (2013–2023). Most works explained are part of a B.Sc., M.Sc., or Ph.D. dissertation.

2.2 AUGMENTED REALITY IN EDUCATION

Due to the COVID-19 pandemic, it was found that traditional learning models were not suitable for teaching from home. Therefore, different technological approaches were tested to encourage the teaching–learning process to be more attractive, motivating, online, and meaningful to the students (Garzón, 2021). One of the key technologies to support the newest teaching–learning strategies is information and communications technologies (ICTs). Within ICTs, AR is widely used for educational purposes because it increases students' learning interest and motivation (Vergara et al., 2016).

Many studies in the state-of-the-art have reported the benefits of using AR in educational settings, including student achievement increase, autonomy facilitation (self-learning), generation of positive attitudes to the educational activity, commitment, motivation, knowledge retention, interaction, collaboration, and availability for all (Lee, 2012; López et al., 2023; Sırakaya & Sırakaya, 2022; Wu et al., 2013). AR in education represents a valuable didactic tool in the teaching–learning process (Kerawalla et al., 2006).

The intention of developing an application based on AR is not to supplant the teacher, but, on the contrary, it allows the instructor of a particular subject, such as

DOI: 10.1201/9781003435198-2

chemistry, Spanish, or history, to improve the methods and tools used in their daily didactics. The combination of mobile devices and AR technology offers advantages such as transportability and virtuality, providing much more extensive real-time information about a place. Students can interact through their mobile devices (tablet, smartphone) with various multimedia elements, such as audio, video, and virtual models, creating an enriched learning experience (Garzón et al., 2019).

The following subsections describe seven works the book's authors have realized in education: history, chemistry, mathematics, language, environmental care, reading, and financial mathematics.

2.2.1 HISTORY

History is not far from technology. Traditional history teaching includes films, photos, news, handouts, and audio. In addition, students can learn about historical sites during school trips. However, the facilities are not always available because resources are often limited, reducing the possibilities for teaching outside the classroom (Evans, 1988). Knowing history is crucial for humans because if the past and its problems are known, the present and its social and political phenomena can be understood.

The B.Sc. computer science dissertation by Rivas and Reyes (2016) developed an AR application to teach students the most relevant characters and events in the history of Mexican independence. A mobile device and a book were employed to narrate and depict the most important events in Mexican independence.

The users read passages about independence from the printed book, and by pointing out the markers with the camera's device, the corresponding scenario is shown through AR. The application was tested with 150 elementary school children. Figure 2.1 shows an example of the students employing the Mexican independence app.

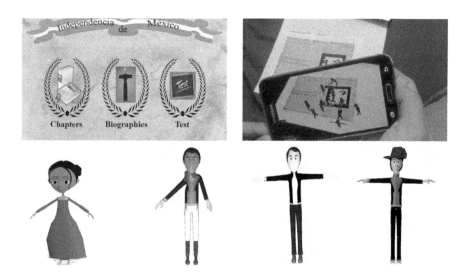

FIGURE 2.1   The Mexican independence AR application.

After testing the app, students expressed that they found AR learning fun and would like similar tools for learning other subjects. Readers are recommended to consult the works of Challenor and Ma (2019) and Remolar et al. (2021) to gain knowledge about the use of AR for history learning.

2.2.2 Chemistry

Chemistry is a science that explains the macroscopic properties of matter from its structure made up of submicroscopic entities called particles. Unfortunately, most students find chemistry courses difficult because they contain complex and abstract information. The subject of chemistry contains concepts and formulas that are taught in a theoretical way. Consequently, the disinterest of students can be generated.

Some students consider chemistry boring, trivial, annoying, and complex. Therefore, it is imperative to propose new ways to teach chemistry (Spencer, 1999). The B.Sc. computer science dissertation of Lara (2016) proposed an AR system to learn about five chemical elements and four compounds. The elements included were carbon, hydrogen, oxygen, sodium, and potassium. In contrast, the compounds included were carbon monoxide, sodium oxide, water, and potassium oxide.

Figure 2.2 shows an example of the AR app. The elements are displayed on the left part, and the compounds on the right. When selecting an element or compound, the user must show the corresponding marker and the 3D model is displayed. This project was tested by 54 students and six teachers from a Mexican public high school.

One hundred percent of teachers and 98% of students liked the app and expressed interest in using it to learn chemistry in the classroom. Students were motivated to learn using a mobile device. Readers can consult the works by Fombona et al. (2022) and Mazzuco et al. (2022) to gain insight into how AR was employed for chemistry education.

FIGURE 2.2   AR application to learn chemistry elements and compounds.

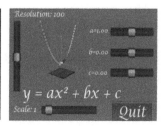

FIGURE 2.3   The pARabola application interface and the fiducial marker.

2.2.3 MATHEMATICS (QUADRATIC EQUATIONS)

Mathematics is a knowledge needed by students to comprehend many real-life concepts. Historically, Mexican students' achievement rate in mathematics has been low (Bosch & Trigueros, 2002). Learning mathematics involves understanding several complex concepts. After conducting a study in three Mexican universities, it was observed that professors have problems teaching the concept of a quadratic equation.

A quadratic equation is an algebraic second-degree equation containing terms with powers no higher than two. The typical form of a quadratic equation is $y = ax^2 + bx + c$, where x is the variable, and a, b, and c are constants. The graph of a quadratic equation is a "U"-shaped curve called a parabola. Therefore, when a teacher explains quadratic equations and the constant changes, a new graph should be drawn to observe the upward or downward effect.

The Ph.D. dissertation by Barraza (2015) presented an AR application called pARabola used to graph a quadratic equation. This application focuses on understanding the concept of the quadratic equation and how its graph representation behaves when its coefficients change. A particle emission system controls the behavior of the graph.

A total of 59 Mexican undergraduate students tested the application. An example of the students experimenting with pARabola and the graphical user interface is shown in Figure 2.3. After the testing stage, almost all the students considered using the pARabola application an excellent way to understand the concept of quadratic equations.

Using pARabola in the classroom created a suitable environment for students to increase their enthusiasm to obtain new knowledge. Moreover, professors considered pARabola an excellent resource for teaching quadratic equations. Demitriadou et al. (2020) and Lai and Cheong (2022) presented reviews about using AR to support mathematics teaching–learning processes.

2.2.4 LANGUAGE

Learning a language different from the mother tongue allows communication and interaction in multilingual communities. Moreover, learning a new language improves memory and brain functioning (Tao & Gao, 2022). Teaching–learning a language is typically developed face-to-face among teachers and students, and great linguistic wealth is involved. Because learning a new language is not easy, new strategies

FIGURE 2.4 User experimenting with the Mayo lottery application.

have been developed to make the task easier. One of the most used strategies is to employ AR. A systematic review regarding the use of AR for language learning can be consulted in the research by Parmaxi and Demetriou (2019).

The Ph.D. work by Miranda et al. (2016) presented a system called "Mayo Lottery" to support the Mayo language learning. Mayo is an indigenous language spoken in the Mexican states of Sonora and Sinaloa. Unfortunately, the number of Mayo speakers is decreasing considerably. Therefore, the AR application aims to teach the language and support its preservation and dissemination.

Mayo Lottery employs images and audio to create a card game. The deck of cards is composed of 89 images. The drawings on each card represent typical objects, animals, and people from the Mayo social group and were used as markers. The game starts when the player selects a board. Then, the cantor randomly selects a card and shows it to the mobile device camera for recognition. As a result, the real model of the object is superimposed in the video stream, and the corresponding audio with the pronunciation of the Mayo word is reproduced, followed by an explanation in Spanish. The process is repeated until the first player completes all the images on the board and shouts Lotería! Figure 2.4 depicts a user experimenting with the application and an example of the board.

The application was tested by 85 undergraduate students from an Indigenous university in Mexico. After using the application, students responded to a survey with 17 items concerning the use and technology acceptance and cultural dimensions of individualism and uncertainty avoidance. The students were satisfied with the operation of the application. However, they recommended that two users should use the app simultaneously to promote collaboration and team learning.

2.2.5 ENVIRONMENTAL CARE

Environmental impact occurs when humans directly or indirectly alter or modify the environment. The consequences of the environmental impact vary depending on the alteration degree. However, alterations can cause diseases, loss of biodiversity, and pollution. Environmental education aims to promote values in people to use natural resources rationally. Moreover, environmental education should be permanent inside and outside schools (Newman, 2006).

Humans are experiencing the consequences of climate change. Therefore, governments worldwide are making great efforts to educate their citizens so that their actions do not have an environmental impact. Simulations can be created using technology aiming to enable people to observe how their habitat could end up if they continue to carry out irresponsible actions toward the environment. AR is one of the technologies that could help educate humans about environmental impact (Cosio et al., 2023).

The B.Sc. computer science dissertation by Tobías (2018) presented an AR application called "AReco" to teach primary school children about environmental care and the consequences of polluting the air, water, and environment. Actions are generated on each of the scenarios depending on the information that is presented through markers. The actions can be positive or negative, depending on which one is selected; the consequences will be reflected in the main scenarios.

Figure 2.5 shows the AReco graphical user interface and water and air contamination examples. One hundred and twenty primary school students tested Areco, and then they filled out a survey to reveal their perception of the application. The students

FIGURE 2.5   AReco graphical user interface and water and air contamination examples.

expressed interest in using AR to learn about environmental care. However, they indicated they liked combining markers the most so that the environment changed accordingly. Other environmental care work can be consulted in Ducasse (2020) and Kamarainen et al. (2013).

2.2.6 READING

Learning to read is a challenge that is often complicated. How quickly and effectively one learns to read depends on the methods used by the teacher. Not all children learn in the same way. Therefore, teachers must use various techniques, such as the global, syllabic, and phonetic methods. Children who learn to read at early stages gain many psychological benefits, including stimulating the mind, delaying cognitive wear, improving concentration, and reducing stress levels (Hargrave & Sénéchal, 2000).

It has been proven that learning to read with AR can bring benefits. Learning to read with AR is usually based on storybooks (Bursali & Yilmaz, 2019). Storybooks are attractive to children because they feel part of the story. An interesting comparison of reading with AR and reading with printed storybooks was presented by Danaei et al. (2020).

The B.Sc. computer science dissertation by Aguilera (2015) presented an AR application with the storybook of the ugly duckling to help children learn to read. Hans Christian Andersen wrote The Ugly Duckling in November 1843. Children like the story because it represents a character who experiences various adversities and was able to fulfill his dream of being happy in the end.

The application aims to arouse students' interest in reading from their early academic education. A printed storybook was employed for reading. On each page, a marker was inserted. The information on each page was brief so as not to bore or tire the children. Therefore, after reading a page, the child must point out the mobile device's camera to observe a 3D model related to the information read, as can be observed in Figure 2.6.

The application was tested by 30 preschool children from a Ciudad Juarez, Mexico college. The application was so successful that the children were motivated to read to see the ugly duckling appear on their work tables. Moreover, the children asked their teachers to prepare more storybooks based on AR.

FIGURE 2.6   The ugly duckling AR application.

2.2.7 FINANCIAL MATHEMATICS (SIMPLE INTEREST)

Financial education must start at an early stage of life. Moreover, financial education helps prevent problems such as having low credit scores or defaulting on a loan (Sun et al., 2020). However, according to Arceo and Villagómez (2017), Mexico reported minimum benefits from including financial education in schools. Therefore, we monitored undergraduates from financial mathematics courses at four public northern Mexican universities to understand why students are not interested in financial education.

From the monitoring, three problems were detected: (i) students lack mathematical skills; (ii) the techniques used by the professors to teach the basics are boring; and (iii) students do not comprehend the basics, such as simple and compound interest, which are fundamental to sound financial education.

The work by Hernández et al. (2021) introduces a prototype called "Simple Interest Computation with Mobile Augmented Reality" (SICMAR) and evaluates its effects on 103 students in a financial mathematics course. A study was designed to assess students' motivation, achievement, technology acceptance, and prototype quality.

Figure 2.7 shows the screen for computing simple interest. First, the user shows a marker representing the desired output, which can be principal, interest rate, amount, simple interest, or time. The prototype waits for the user to show the markers for input terms. Then, verification is conducted to determine if the necessary data for the computation were inserted correctly. Finally, the result of the calculations is displayed.

Students considered that SICMAR could be applied to financial mathematics, obtaining the benefit of increasing perception and user interaction with the environment; a non-AR application cannot offer those features. MAR changes how students interact with the world. Moreover, using SICMAR increases students' motivation

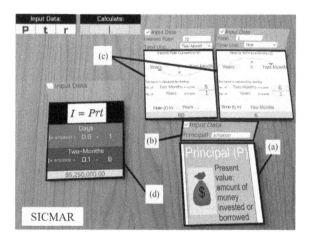

a) Display information.

b) Capture user inputs.

c) Time conversions

d) Results/errors.

FIGURE 2.7   The screen for simple interest computation with mobile AR.

when learning about simple interest topics. More examples of using AR in finance education were proposed by Candra et al. (2022) and Maad et al. (2008).

In summary, the educational field is one of the areas where AR has begun to have more remarkable development. Thanks to technology, science teaching strategies and methodologies have been reconsidered. Nowadays, students are expected to be able to reason and be creative and innovative in solving problems in the development area that concerns them. Therefore, it can be asserted that AR results in enjoyment in students and a desire to repeat the experience.

2.3 AUGMENTED REALITY IN MEDICINE

Medicine is a science that focuses on developing techniques to treat and prevent diseases to preserve people's health. Nowadays, specialized medical equipment can generate diagnoses or surgeries with minimal error (Yuan et al., 2016). Therefore, thanks to technology, precise, fast, and less invasive techniques have been developed in medicine. However, the clinical implementations of technologies such as artificial intelligence and AR have not yet become a reality (He et al., 2019).

AR in medicine is undoubtedly valuable and helpful to both doctors and patients. AR could support doctors during surgery, improve training, or visualize data. Moreover, doctors can employ AR to explain to a patient about a disease better. Various medical specialties have benefited from using AR, for example, dental medicine (Joda et al., 2019), anatomy (Duarte et al., 2020), and surgery (Vávra et al., 2017).

The following subsections describe four works that the book's authors have realized in medicine: vital signs monitoring, phobia treatments, medication adherence, and color blindness.

2.3.1 VITAL SIGNS MONITORING

In medicine, the accurate measuring of vital signs is important because they inform how well the body works. Vital signs include heart rate, respiratory rate, temperature, and blood pressure, and aid in disease prevention (Edmonds et al., 2002). Unfortunately, nurses in some hospitals must care for many patients and spend much time keeping up-to-date vital sign information records for each individual. Therefore, using an intelligent monitoring system could be an ideal solution to solving this problem.

The M.Sc. work by González et al. (2014) proposed a real-time smart multilevel tool for remote patient monitoring of the body temperature and heart rate. The proposal's core is a wireless sensor network (WSN) and MAR. WiFi technology is employed to transmit data, reports, and alarms to a nurse server or a mobile device in wireless mode. On the other hand, smart nodes evaluate the patient's condition through data obtained from biometric sensors (body temperature and heart rate). Finally, AR virtual patient files are implemented to complement the handwritten nursing reports. Temperature measurements detect hyperthermia, hypothermia, or average body temperature. Additionally, the heartbeat sensor was able to detect tachycardia or bradycardia conditions.

FIGURE 2.8   AR app for monitoring a patient's body temperature and heart rate.

As shown in Figure 2.8, the tool uses two markers to display the body temperature and heart rate. Both markers are located on the patient's headboard. When the nurse approaches a bed, she executes the application. Instantly, live video from the mobile device camera is acquired, and the nurse must point to the markers, which are recognized using computer vision techniques. A 2D thermometer image and measured temperature value (in Celsius) are superimposed onto the video stream if the tool detects the body temperature marker. A 2D heart image and the measured value (beats per minute) are superimposed if the heart rate marker is recognized.

All the data obtained are stored in a database to maintain information about each patient. The test results show that the system performs effectively within a range of 20 meters and requires 10 minutes to stabilize the temperature sensor. Other works that employed AR for patient monitoring were published by Arpaia et al. (2021) and Plabst et al. (2021).

2.3.2   PHOBIA TREATMENTS

A phobia is a persistent, excessive, and irrational fear provoked by the presence of a specific situation or object. Moreover, phobias are the most common anxiety disorder. Phobias are divided into four types, including: (i) situational, (ii) natural environment, (iii) animal, and (iv) blood/injection/injury. Approximately 5% of the world's population suffers from a phobia (Wolitzky et al., 2008).

Many people avoid treatment for their phobia due to ignorance of the therapeutic procedures. There are different therapies to treat phobias, such as live exposure therapy and cognitive behavioral therapy. In live exposure therapy, the patient confronts his or her fear systematically and deliberately. On the contrary, cognitive behavioral therapy involves gradual exposure and other ways to learn how to cope with the feared object (Grös & Antony, 2006).

The B.Sc. computer science dissertation by Hernández and Vázquez (2016) presented an AR application to support live exposure therapy to treat arachnophobia. The system confronts the patient with their fear of spiders with the therapist's help. Among the advantages of the system over traditional therapy is that there is no need for a spider. Therefore, there is no risk of a bite. Figure 2.9 shows an example of the system in operation.

FIGURE 2.9   AR app for arachnophobia treatment.

Ten users tested the system and mentioned that it helped their therapy process. In addition, therapists have also expressed interest in using the tool in their practice. Readers can consult the works by Albakri et al. (2022) and Chicchi et al. (2015) to obtain a vast panorama regarding the use of AR for phobia treatment.

2.3.3 MEDICATION ADHERENCE

According to Mohan et al. (2018), there are three main elements to treating a disease: (i) a correct diagnosis, (ii) medications available, and (iii) treatments must be followed according to the medical prescription, which is known as medication adherence. Therefore, taking the appropriate doses of medicine at defined times is a recommended practice to cure or overcome disease. Interrupting a treatment or not taking the medication at defined times can have severe consequences on a person's health and even cause death. Unfortunately, older adults are vulnerable to disease and often do not have control over adhering to short- or long-term medical treatment.

Some adults do not adhere to their treatment due to voluntary or involuntary forgetfulness, because they are confused about the time to take each dose, or because they suffer from an illness that prevents them from remembering the medication schedule. Therefore, technology can help adults adhere to their medical treatment. A review regarding the use of technologies for medication adherence can be consulted in the manuscript by Aldeer et al. (2018).

The Bs.C. computer science dissertation by De la Torre and López (2018) proposed MedicAR, an application based on optical character recognition (OCR) and AR to help people with medication adherence. Figure 2.10 shows that MedicAR recognizes numerical codes pasted in each medicine. Then, a query to a database is conducted to obtain information regarding the frequency and times at which the medication should be taken. In addition, the information in the database is used to send reminders to the patient.

FIGURE 2.10 The information displayed by MedicAR.

The characteristics of MedicAR were determined through a survey applied to 100 users. In addition, 10 older adults tested MedicAR during their treatment and expressed satisfaction with the application. However, we observed that using technology can cause discomfort for some older people.

2.3.4 COLOR BLINDNESS

Color blindness is a visual impairment that results in the inhibition of color perception in people. Humans with color blindness usually have a narrower perception of the color spectrum than people with normal color vision due to a genetic anomaly of the cones in the eye. Color blindness can be acquired or congenital. The most common forms of color blindness are protanopia, where red is not perceived, and deuteranopia, where green is not perceived (Wong, 2011).

Although some people consider color blindness not to be a significant condition, it can cause problems in the development of many activities. For example, if a person cannot detect whether a traffic light is green or red, an accident can be caused. Therefore, since there is no cure for color blindness and it cannot be corrected with eyeglasses or contact lenses, technology can be employed to mitigate it (Salih et al., 2020).

AR is one of the technologies that can be used to support people with color blindness (Tanuwidjaja et al., 2014). Therefore, the Bs.C. mechatronics dissertation by Gálvez (2017) proposed an AR system to support people with deuteranopia-type color blindness. The system uses a test with 32 graphic questions to determine the colors with which the colorblind person has the most problems. The system can be proved if the analysis determines the person has deuteranopia by evaluating the questions.

The user must determine the object's color and the background color of the image target shown to the system. Some users cannot determine the figure's color, while others cannot detect the figure. Hence, the AR system employs the mobile device's camera to help users recognize colors using audio. Moreover, a color strip is superimposed in the video stream in which the unrecognized color is highlighted using a color that the user can see. Figure 2.11 shows a person using the AR system.

Ten persons with deuteranopia-type color blindness tested the system. The results revealed that the system can help a person identify colors and differentiate between two colors in which previously they could only see one. Readers who want to obtain more information regarding the use of AR in ophthalmology can consult the work by Aydindoğan et al. (2021).

FIGURE 2.11   AR application for color blindness.

In summary, nowadays, and as ever, technology and medicine are very interrelated. However, the insertion of AR into operating or hospital rooms must be legalized. Moreover, the lack of funding for product development, technical limitations, and confidentiality in data management are barriers that must be overcome. Even so, AR may be the future of personalized care.

2.4   AUGMENTED REALITY IN INDUSTRY

Day by day, industries such as manufacturing, electronics, food, automotive, textile, and construction, to mention a few, work to improve their operations, increase revenues, and transform products aiming for customer satisfaction. Companies are beginning to include technologies such as robotics, artificial intelligence, cognitive computing, data science, the Internet of Things, and virtual and augmented reality in their operations. Due to the insertion of technology in industry, what is known as Industry 5.0 has been created. The aim of Industry 5.0 is that human experts collaborate with machines to create products most efficiently (Mukherjee et al., 2023).

AR plays an essential role in the transition toward Industry 5.0. Product design, manufacturing, assembly, maintenance, inspection, and training are core areas of the industry to which AR can contribute. Consequently, Fite (2011) coined the term "industrial augmented reality" to describe the use of AR for supporting industrial processes. Interested readers should consult the works of Bottani and Vignali (2019), de Souza et al. (2020), and Lavingia and Tanwar (2020) to obtain an extended view of the use of AR in industry.

In developed countries such as Germany, China, and Japan, the philosophies related to Industry 5.0 have been successfully proven and adopted. However, in other regions, such as Mexico, adopting the philosophies associated with Industry 5.0 has been slightly delayed. Therefore, the following subsections depict the efforts conducted by this book's authors to support industries by implementing AR in their production processes.

2.4.1   Manufacturing of All-Terrain Vehicles

The automotive industry is related to motor vehicle design, development, manufacturing, marketing, and selling. All-terrain vehicles (ATVs) are a branch of the automotive industry introduced in the United States in 1971 (Schöner, 2004). Aesthetic and functional changes in the design of the existing ATVs happen frequently. All these changes are subjected to rigorous quality controls and must comply with strict security measures. Therefore, training personnel who design and manufacture ATVs is complex and needs to be done rapidly.

The Bs.C. mechatronics dissertation by Nava (2017) presented a mobile AR prototype for helping manufacture an ATV. The prototype comprises three operations: (i) welding inspection, (ii) measuring of critical dimensions, and (iii) mounting of virtual accessories in the ATV chassis. Welding inspection verifies the welding features against the specifications defined by the quality department. As observed in Figure 2.12, information about the weld number assigned, the workstation (cell) location where

FIGURE 2.12   AR scenes for manufacturing ATVs.

the weld was made, the importance of the weld, and the weld trajectory type are superimposed on the scene. The AR tool for measuring critical dimensions offers a guide to checking the measures that impact the quality of a product. The dimensions from one component to another and dimensions from individual components were included. Finally, the experience shows the place in the chassis where the accessories will be mounted.

Ten employees of the ATV manufacturing company filled out a questionnaire to measure user satisfaction when using the application. The results demonstrated that the prototype is useful for manufacturing an ATV, including the training stage. The

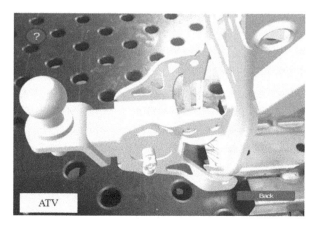

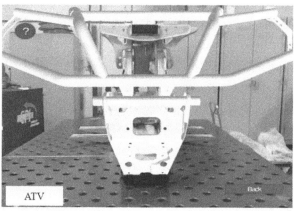

FIGURE 2.12 (Continued)

manuscript by Boboc et al. (2020) can be consulted to learn how AR has been used in the automotive industry.

2.4.2 MACHINING TRAINING

Training is the set of procedures and activities conducted to increase a person's abilities. In industries, qualified professionals are needed. Therefore, training has a direct impact on productivity. Even though the benefits of training are clear, there are still barriers that must be overcome, such as costs and worker resistance, and they cannot last for extended periods (Mitchell et al., 2004).

Classroom style is the traditional training method in the industry. Presentation methods and hands-on training are also employed frequently. However, simulation is the best training option when skills for operating complex machines are needed. The simulator must recreate easy to complex situations, aiming for trainees to solve almost all situations in real environments. Compared to classical training approaches,

FIGURE 2.13   The AR scenes and markers employed for lathe and milling training.

simulators allow employees to operate equipment before it has been installed. Virtual and augmented reality can be recommended tools for developing virtual environments for training (Radhakrishnan et al., 2021; Webel et al., 2013).

The Bs.C. mechatronics dissertation by Monroy (2015) proposes a MAR system to train workers using milling and lathe machines. The system incorporates 3D models of machinery and tools, text instructions, animations, and videos with real processes. Moreover, tutorials on performing the essential tasks of tools setup, setup of working material, and machinery setup and start-up for milling and a lathe machine were included. Figure 2.13 shows scenes of the AR application and examples of the markers used.

Since using both hands is imperative for operating a lathe and milling, the ORA-1 AR glasses and the VR-PRO AR HMD were employed to display the AR experience. Sixteen workers rated the system as a valuable and appealing tool for learning and

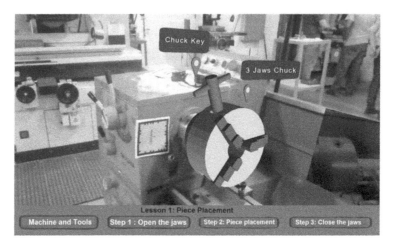

FIGURE 2.13 (Continued)

training the basics of lathe and milling manipulation. Moreover, the system represents a real possibility to expand the machinery handling support offered by the instructor.

2.4.3 Computer Maintenance

Ensuring productivity is an imperative task in the industry. Therefore, all equipment types inside the manufacturing plant must always be guaranteed to work correctly. Maintenance is defined as activities conducted to correct equipment failures and ensure proper operation. Mainly, there are three maintenance types: corrective, predictive, and preventive. Corrective maintenance is executed immediately after a defect is detected in a piece of equipment so that it can function normally again. Predictive maintenance predicts when an equipment failure could occur and thus be able to avoid it. Preventive maintenance is performed regularly following a schedule

FIGURE 2.14 The instructions and 3D models for processor maintenance and keyboard cleaning.

regardless of the equipment's condition (Wang et al., 2020; Yepez et al., 2019; Zonta et al., 2020).

In industrial settings, computers are used to control, monitor, correct, and improve processes to ensure the manufacture of quality products. The maintenance of a computer is executed to ensure its proper functioning. In computer maintenance, parts are cleaned, viruses are removed, antivirus is updated, and software and hardware updates are carried out (Lewis, 1964).

The Bs.C. computer science dissertation by Ortega (2019) developed a markerless AR prototype for the preventive and corrective maintenance of personal computers (PCs). The prototype uses 3D animated models and text instructions to explain the steps to conduct maintenance. Moreover, the prototype explains the tools to be used and the care that must be taken to avoid damaging components. When the user points the camera at the computer motherboard, the labels corresponding to the processor, heat sink, RAM, graphics card, hard disk, and power supply are overlaid in the video stream. In addition, the interface has a menu that allows users to select the component's animation that will be maintained. Figure 2.14 shows an example of the instructions for maintaining the processor and keyboard cleaning.

Fifteen users completed the satisfaction survey. All the users expressed feeling motivated when using the application. However, they considered the number of tasks too small and requested that more activities be included. Readers can consult the proposals by Kang et al. (2016) and Westerfield et al. (2015) for other examples regarding computer maintenance.

2.4.4 PLAN INTERPRETATION AND VISUALIZATION

At various stages of the manufacturing process, products must be inspected to ensure their quality. The task is called the quality inspection process (QIP). The quality inspector must review the product plan, which includes different views and the measurements of each part. However, the plan can contain many references, so reading and interpreting it can be complex and time-consuming (Babic et al., 2021).

The Bs.C. mechatronics dissertation by Beltrán (2016) developed an AR tool that employs multiple markers to support plan interpretation and visualization. The

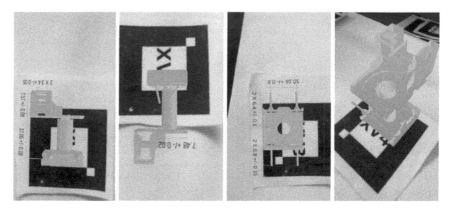

FIGURE 2.15 The marker, the virtual model, and the critical dimensions for plan interpretation.

prototype provides information on the number of physical parts to be measured, the critical dimensions and tolerances that affect production, and the measuring instrument that needs to be employed (vernier or micrometer).

Figure 2.15 shows an example of the marker used, the virtual model of a relay spool, and critical dimension information. The different views of the virtual model were created with AutoCAD. Moreover, 3D text was inserted into the scene to display the critical dimensions.

Two quality inspectors were asked to conduct the QIP for 10 different part numbers with and without using the AR prototype. The analysis was divided into plans with more than 100 and less than 100 dimensions. The average time to complete the task in plans with less than 100 dimensions without using AR was 19 minutes, and with AR was 10 minutes. For plans with more than 100 dimensions, the task took 28 minutes without AR and 18 minutes with AR.

In summary, today's industries must customize their products according to customer needs. In addition, the delivery period of the products must be short, and competitive costs must also be offered. One alternative to accomplish this implies adding technology to the production lines. According to de Souza et al. (2020), AR improves industries' flexibility and process efficiency. Unfortunately, the implementation of AR in industries is still not standardized.

REFERENCES

Aguilera, M. (2015). *Aplicación Móvil como Apoyo para el Fomento de la Lectura Infantil Utilizando Realidad Aumentada* [B.Sc. dissertation]. Universidad Autonoma de Ciudad Juarez.

Albakri, G., Bouaziz, R., Alharthi, W., Kammoun, S., Al-Sarem, M., Saeed, F., & Hadwan, M. (2022). Phobia Exposure Therapy Using Virtual and Augmented Reality: A Systematic Review. *Applied Sciences*, *12*(3), 1–20. https://doi.org/10.3390/app12031672

Aldeer, M., Javanmard, M., & Martin, R. (2018). A Review of Medication Adherence Monitoring Technologies. *Applied System Innovation*, *1*(2), 1–26. https://doi.org/10.3390/asi1020014

Arceo, E., & Villagómez, F. (2017). Financial Literacy Among Mexican High School Teenagers. *International Review of Economics Education*, *24*, 1–17. https://doi.org/https://doi.org/10.1016/j.iree.2016.10.001

Arpaia, P., De Benedetto, E., Dodaro, C., Duraccio, L., & Servillo, G. (2021). Metrology-Based Design of a Wearable Augmented Reality System for Monitoring Patient's Vitals in Real Time. *IEEE Sensors Journal*, *21*(9), 11176–11183. https://doi.org/10.1109/JSEN.2021.3059636

Aydindoğan, G., Kavakli, K., Şahin, A., Artal, P., & Ürey, H. (2021). Applications of Augmented Reality in Ophthalmology. *Biomedical Optics Express*, *12*(1), 511–538. https://doi.org/10.1364/BOE.405026

Azuma, R., Baillot, Y., Behringer, R., Feiner, S., Julier, S., & MacIntyre, B. (2001). Recent Advances in Augmented Reality. *IEEE Computer Graphics and Applications*, *21*(6), 34–47. https://doi.org/10.1109/38.963459

Babic, M., Farahani, M., & Wuest, T. (2021). Image Based Quality Inspection in Smart Manufacturing Systems: A Literature Review. *Procedia CIRP*, *103*, 262–267. https://doi.org/https://doi.org/10.1016/j.procir.2021.10.042

Barraza, R. (2015). *Arquitectura de Software para el Diseño de Material Didáctico Basado en Realidad Aumentada Móvil Colaborativa* [Ph.D. dissertation]. Universidad Autónoma de Ciudad Juárez.

Beltrán, Á. (2016). *Realidad Aumentada como Herramienta para Interpretación de Planos* [B.Sc. dissertation]. Universidad Autonoma de Ciudad Juarez.

Boboc, R., Gîrbacia, F., & Butilă, E. (2020). The Application of Augmented Reality in the Automotive Industry: A Systematic Literature Review. *Applied Sciences*, *10*(12), 1–22. https://doi.org/10.3390/app10124259

Bosch, C., & Trigueros, M. (2002). Gender and Mathematics in Mexico. In Hanna, G. (Ed.), *Towards Gender Equity in Mathematics Education: An ICMI Study* (pp. 277–284). Springer Netherlands. https://doi.org/10.1007/0-306-47205-8_19

Bottani, E., & Vignali, G. (2019). Augmented Reality Technology in the Manufacturing Industry: A Review of the Last Decade. *IISE Transactions*, *51*(3), 284–310.

Bursali, H., & Yilmaz, R. (2019). Effect of Augmented Reality Applications on Secondary School Students' Reading Comprehension and Learning Permanency. *Computers in Human Behavior*, *95*, 126–135. https://doi.org/https://doi.org/10.1016/j.chb.2019.01.035

Candra, R., Rika, P., Ilyana, S., & Dwi, H. (2022). Augmented Reality (AR)-Based Sharia Financial Literacy System (AR-SFLS): A New Approach to Virtual Sharia Financial Socialization for Young Learners. *International Journal of Islamic and Middle Eastern Finance and Management*, *15*(1), 48–65. https://doi.org/10.1108/IMEFM-11-2019-0484

Challenor, J., & Ma, M. (2019). A Review of Augmented Reality Applications for History Education and Heritage Visualisation. *Multimodal Technologies and Interaction*, *3*(2), 1–20. https://doi.org/10.3390/mti3020039

Chicchi, I., Pallavicini, F., Pedroli, E., Serino, S., & Riva, G. (2015). Augmented Reality: A Brand New Challenge for the Assessment and Treatment of Psychological Disorders. *Computational and Mathematical Methods in Medicine*, *2015*, 1–12. https://doi.org/10.1155/2015/862942

Cosio, L., Buruk, O., Fernández, D., De Villiers, I., & Hamari, J. (2023). Virtual and Augmented Reality for Environmental Sustainability: A Systematic Review. *Proceedings of the CHI Conference on Human Factors in Computing Systems*, pp. 1–23. https://doi.org/10.1145/3544548.3581147

Danaei, D., Jamali, H., Mansourian, Y., & Rastegarpour, H. (2020). Comparing Reading Comprehension Between Children Reading Augmented Reality and Print Storybooks. *Computers & Education*, *153*, 1–10. https://doi.org/https://doi.org/10.1016/j.compedu.2020.103900

De la Torre, L., & López, E. (2018). *Prototipo de Aplicación Móvil con Realidad Aumentada como Apoyo en la Adherencia a un Tratamiento Farmacológico* [B.Sc. dissertation]. Universidad Autonoma de Ciudad Juarez.

Demitriadou, E., Stavroulia, K.-E., & Lanitis, A. (2020). Comparative Evaluation of Virtual and Augmented Reality for Teaching Mathematics in Primary Education. *Education and Information Technologies*, *25*(1), 381–401. https://doi.org/10.1007/s10639-019-09973-5

de Souza, L., Martins, F., & Zorzal, E. (2020). A Survey of Industrial Augmented Reality. *Computers & Industrial Engineering*, *139*, 1–14. https://doi.org/https://doi.org/10.1016/j.cie.2019.106159

Duarte, M., Santos, L., Guimarães, J., & Peccin, M. S. (2020). Learning Anatomy by Virtual Reality and Augmented Reality. A Scope Review. *Morphologie*, *104*(347), 254–266. https://doi.org/https://doi.org/10.1016/j.morpho.2020.08.004

Ducasse, J. (2020). Augmented Reality for Outdoor Environmental Education. In Geroimenko, V. (Ed.), *Augmented Reality in Education: A New Technology for Teaching and Learning* (pp. 329–352). Springer International Publishing. https://doi.org/10.1007/978-3-030-42156-4_17

Edmonds, Z., Mower, W., Lovato, L., & Lomeli, R. (2002). The Reliability of Vital Sign Measurements. *Annals of Emergency Medicine*, *39*(3), 233–237. https://doi.org/https://doi.org/10.1067/mem.2002.122017

Evans, R. (1988). Lessons from History: Teacher and Student Conceptions of the Meaning of History. *Theory & Research in Social Education*, *16*(3), 203–225. https://doi.org/10.1080/00933104.1988.10505565

Fite, P. (2011). Is there a reality in Industrial Augmented Reality? *Proceedings of the 10th IEEE International Symposium on Mixed and Augmented Reality (ISMAR)*, 201–210. https://doi.org/10.1109/ISMAR.2011.6092387

Fombona, A., Fombona, J., Vicente, R. (2022) Augmented Reality, a Review of a Way to Represent and Manipulate 3D Chemical Structures. *Journal of Chemical Information and Modeling* 62(8), 1863–1872.

Gálvez, J. (2017). *Sistema de Realidad Aumentada Para Apoyar a Personas con Daltonismo de Tipo Deuteranopia* [B.Sc. dissertation]. Universidad Autonoma de Ciudad Juarez.

Garzón, J. (2021). An Overview of Twenty-Five Years of Augmented Reality in Education. *Multimodal Technologies and Interaction*, *5*(7), 1–14. https://doi.org/10.3390/mti5070037

Garzón, J., Pavón, J., & Baldiris, S. (2019). Systematic Review and Meta-Analysis of Augmented Reality in Educational Settings. *Virtual Reality*, *23*(4), 447–459. https://doi.org/10.1007/s10055-019-00379-9

González, F., Vergara, O., Ramírez, D., Cruz, V., & Ochoa, H. (2014). Smart Multi-Level Tool for Remote Patient Monitoring Based on a Wireless Sensor Network and Mobile Augmented Reality. *Sensors*, *14*(9), 17212–17234. https://doi.org/10.3390/s140917212

Grös, D., & Antony, M. (2006). The Assessment and Treatment of Specific Phobias: A Review. *Current Psychiatry Reports*, *8*(4), 298–303. https://doi.org/10.1007/s11920-006-0066-3

Hargrave, A., & Sénéchal, M. (2000). A Book Reading Intervention with Preschool Children who Have Limited Vocabularies: The Benefits of Regular Reading and Dialogic Reading. *Early Childhood Research Quarterly*, *15*(1), 75–90. https://doi.org/https://doi.org/10.1016/S0885-2006(99)00038-1

He, J., Baxter, S., Xu, J., Xu, J., Zhou, X., & Zhang, K. (2019). The Practical Implementation of Artificial Intelligence Technologies in Medicine. *Nature Medicine*, *25*(1), 30–36. https://doi.org/10.1038/s41591-018-0307-0

Hernández, C., & Vázquez, M. (2016). *Sistema de Realidad Aumentada Mediante Marcadores como Apoyo para el Tratamiento de Aracnofobia* [B.Sc. dissertation]. Universidad Autonoma de Ciudad Juarez.

Hernández, L., López, J., Tovar, M., Vergara, O., & Cruz, V. (2021). Effects of Using Mobile Augmented Reality for Simple Interest Computation in a Financial Mathematics Course. *PeerJ Computer Science, 7:e618*(1), 1–33. https://doi.org/10.7717/peerj-cs.618

Joda, T., Gallucci, G., Wismeijer, D., & Zitzmann, N. (2019). Augmented and Virtual Reality in Dental Medicine: A Systematic Review. *Computers in Biology and Medicine, 108*, 93–100. https://doi.org/https://doi.org/10.1016/j.compbiomed.2019.03.012

Kamarainen, A., Metcalf, S., Grotzer, T., Browne, A., Mazzuca, D., Tutwiler, M., & Dede, C. (2013). EcoMOBILE: Integrating Augmented Reality and Probeware with Environmental Education Field Trips. *Computers & Education, 68*, 545–556. https://doi.org/https://doi.org/10.1016/j.compedu.2013.02.018

Kang, B., Ren, P. & Ke, C. (2006). An Augmented Reality System for Computer Maintenance. In Pan, Z., Cheok, A., Haller, M., Lau, R., Saito, H., & Liang, R. (Eds.), *Advances in Artificial Reality and Tele-Existence* (pp. 284–291). Springer Berlin Heidelberg.

Kerawalla, L., Luckin, R., Seljeflot, S., & Woolard, A. (2006). "Making it Real": Exploring the Potential of Augmented Reality for Teaching Primary School Science. *Virtual Reality, 10*(3), 163–174. https://doi.org/10.1007/s10055-006-0036-4

Lai, J., & Cheong, K. (2022). Adoption of Virtual and Augmented Reality for Mathematics Education: A Scoping Review. *IEEE Access, 10*, 13693–13703. https://doi.org/10.1109/ACCESS.2022.3145991

Lara, A. (2016). *Sistema de Realidad Aumentada como Soporte a la Enseñanza de Elementos Químicos* [B.Sc. dissertation]. Universidad Autonoma de Ciudad Juarez.

Lavingia, K., & Tanwar, S. (2020). Augmented Reality and Industry 4.0. In Nayyar, A. & Kumar (Eds.), *A Roadmap to Industry 4.0: Smart Production, Sharp Business and Sustainable Development* (pp. 143–155). Springer International Publishing. https://doi.org/10.1007/978-3-030-14544-6_8

Lee, K. (2012). Augmented Reality in Education and Training. *TechTrends, 56*(2), 13–21. https://doi.org/10.1007/s11528-012-0559-3

Lewis, P. (1964). Implications of a Failure Model for the Use and Maintenance of Computers. *Journal of Applied Probability, 1*(2), 347–368. https://doi.org/DOI: 10.2307/3211865

López, J., Moreno, A., López, J., & Hinojo, F. (2023). Augmented Reality in Education. A Scientific Mapping in Web of Science. *Interactive Learning Environments, 31*(4), 1860–1874. https://doi.org/10.1080/10494820.2020.1859546

Maad, S., Garbaya, S., & Bouakaz, S. (2008). From Virtual to Augmented Reality in Financial Trading: A CYBERII Application. *Journal of Enterprise Information Management, 21*(1), 71–80. https://doi.org/10.1108/17410390810842264

Miranda, E., Vergara, O., Cruz, V., García, J., & Favela, J. (2016). Study on Mobile Augmented Reality Adoption for Mayo Language Learning. *Mobile Information Systems, 2016*, 1069581. https://doi.org/10.1155/2016/1069581

Mitchell, C., Doyle, M., Moran, J., Lippy, B., Hughes Jr., J., Lum, M., & Agnew, J. (2004). Worker Training for New Threats: A Proposed Framework. *American Journal of Industrial Medicine, 46*(5), 423–431. https://doi.org/https://doi.org/10.1002/ajim.20091

Mohan, P., Shaji, S., Ashraf, T., Anas, V., Basheer, B. (2018) Effectiveness of a reminder card system versus a mobile application to improve medication adherence among asthma patients in a tertiary care hospital. *Journal of Taibah University Medical Sciences, 13*(6), 541–546.

Monroy, A. (2015). *Desarrollo de un Sistema de Realidad Aumentada como Herramienta de Apoyo en el Uso de Maquinaria* [B. Sc. dissertation]. Universidad Autonoma de Ciudad Juarez.

Mukherjee, A., Raj, A., & Aggarwal, S. (2023). Identification of Barriers and their Mitigation Strategies for Industry 5.0 Implementation in Emerging Economies. *International Journal of Production Economics, 257*, 108770. https://doi.org/https://doi.org/10.1016/j.ijpe.2023.108770

Nava, E. (2017). *Desarrollo de una Aplicación de Realidad Aumentada Móvil para Entornos de Manufactura en la Industria* [B. Sc. dissertation]. Universidad Autonoma de Ciudad Juarez.

Newman, P. (2006). The Environmental Impact of Cities. *Environment and Urbanization*, *18*(2), 275–295. https://doi.org/10.1177/0956247806069599

Ortega, Y. (2019). *Aplicación Móvil de Realidad Aumentada para el Mantenimiento de Equipo de Cómputo* [B. Sc. dissertation]. Universidad Autonoma de Ciudad Juarez.

Parmaxi, A., & Demetriou, A. (2019). Augmented Reality in Language Learning: A State-of-the-Art Review of 2014–2019. *Journal of Computer Assisted Learning*, *36*, 861–875. https://api.semanticscholar.org/CorpusID:225307291

Plabst, L., Oberdörfer, S., Happel, O., & Niebling, F. (2021). Visualization Methods for Patient Monitoring in Anaesthetic Procedures Using Augmented Reality. *Proceedings of the 27th ACM Symposium on Virtual Reality Software and Technology (VRST)*, pp. 1–3. https://doi.org/10.1145/3489849.3489908

Radhakrishnan, U., Koumaditis, K., & Chinello, F. (2021). A Systematic Review of Immersive Virtual Reality for Industrial Skills Training. *Behaviour & Information Technology*, *40*(12), 1310–1339. https://doi.org/10.1080/0144929X.2021.1954693

Remolar, I., Rebollo, C., & Fernández, J. (2021). Learning History Using Virtual and Augmented Reality. *Computers*, *10*(11), 1–19. https://doi.org/10.3390/computers10110146

Rivas, A., & Reyes, J. (2016). *Software Educativo como Herramienta de Apoyo para la Enseñanza de la Independencia de México Basado en Realidad Aumentada* [B. Sc. dissertation]. Universidad Autonoma de Ciudad Juarez.

Salih, A., Elsherif, M., Ali, M., Vahdati, N., Yetisen, A., & Butt, H. (2020). Ophthalmic Wearable Devices for Color Blindness Management. *Advanced Materials Technologies*, *5*(8), 1–13. https://doi.org/https://doi.org/10.1002/admt.201901134

Schöner, H. (2004). Automotive Mechatronics. *Control Engineering Practice*, *12*(11), 1343–1351. https://doi.org/https://doi.org/10.1016/j.conengprac.2003.10.004

Sırakaya, M., & Sırakaya, D. (2022). Augmented Reality in STEM Education: A Systematic Review. *Interactive Learning Environments*, *30*(8), 1556–1569. https://doi.org/10.1080/10494820.2020.1722713

Spencer, J. (1999). New Directions in Teaching Chemistry: A Philosophical and Pedagogical Basis. *Journal of Chemical Education*, *76*(4), 566. https://doi.org/10.1021/ed076p566

Sun, H., Yuen, D., Zhang, J., & Zhang, X. (2020). Is knowledge Powerful? Evidence from Financial Education and Earnings Quality. *Research in International Business and Finance*, *52*, 101179. https://doi.org/https://doi.org/10.1016/j.ribaf.2019.101179

Tanuwidjaja, E., Huynh, D., Koa, K., Nguyen, C., Shao, C., Torbett, P., Emmenegger, C., & Weibel, N. (2014). Chroma: A Wearable Augmented-Reality Solution for Color Blindness. *Proceedings of the ACM International Joint Conference on Pervasive and Ubiquitous Computing*, 799–810. https://doi.org/10.1145/2632048.2632091

Tao, J., & Gao, X. (2022). Teaching and Learning Languages Online: Challenges and Responses. *System*, *107*, 102819. https://doi.org/https://doi.org/10.1016/j.system.2022.102819

Tobías, D. (2018). *Prototipo Móvil de Realidad Aumentada para Fomentar el Cuidado del Medio Ambiente* [B. Sc. dissertation]. Universidad Autonoma de Ciudad Juarez.

Vávra, P., Roman, J., Zonča, P., Ihnát, P., Němec, M., Kumar, J., Habib, N., & El-Gendi, A. (2017). Recent Development of Augmented Reality in Surgery: A Review. *Journal of Healthcare Engineering*, *2017*, 4574172. https://doi.org/10.1155/2017/4574172

Vergara, O., Cruz, V., Rodríguez, R., & Nandayapa, M. (2016). Recent ADvances in Augmented Reality (RADAR). *International Journal of Combinatorial Optimization Problems and Informatics*, *7*(3), 1–6.

Wang, N., Ren, S., Liu, Y., Yang, M., Wang, J., & Huisingh, D. (2020). An Active Preventive Maintenance Approach of Complex Equipment Based on a Novel Product-Service System Operation Mode. *Journal of Cleaner Production, 277*, 123365. https://doi.org/ https://doi.org/10.1016/j.jclepro.2020.123365

Webel, S., Bockholt, U., Engelke, T., Gavish, N., Olbrich, M., & Preusche, C. (2013). An Augmented Reality Training Platform for Assembly and Maintenance Skills. *Robotics and Autonomous Systems, 61*(4), 398–403. https://doi.org/https://doi.org/10.1016/ j.robot.2012.09.013

Westerfield, G., Mitrovic, A., & Billinghurst, M. (2015). Intelligent Augmented Reality Training for Motherboard Assembly. *International Journal of Artificial Intelligence in Education, 25*(1), 157–172. https://doi.org/10.1007/s40593-014-0032-x

Wolitzky, K., Horowitz, J., Powers, M., & Telch, M. (2008). Psychological Approaches in the Treatment of Specific Phobias: A Meta-Analysis. *Clinical Psychology Review, 28*(6), 1021–1037. https://doi.org/https://doi.org/10.1016/j.cpr.2008.02.007

Wong, B. (2011). Points of View: Color Blindness. *Nature Methods, 8*(6), 441. https://doi.org/ 10.1038/nmeth.1618

Wu, H., Lee, S., Chang, H., & Liang, J. (2013). Current Status, Opportunities and Challenges of Augmented Reality in Education. *Computers & Education, 62*, 41–49. https://doi.org/ https://doi.org/10.1016/j.compedu.2012.10.024

Yepez, P., Alsayyed, B., & Ahmad, R. (2019). Intelligent Assisted Maintenance Plan Generation for Corrective Maintenance. *Manufacturing Letters, 21*, 7–11. https://doi.org/https://doi. org/10.1016/j.mfglet.2019.06.004

Yuan, H., Ma, Q., Ye, L., & Piao, G. (2016). The Traditional Medicine and Modern Medicine from Natural Products. *Molecules, 21*(5), 1–18. https://doi.org/10.3390/molecules2 1050559

Zonta, T., da Costa, C., da Rosa, R., de Lima, M., da Trindade, E., & Li, G. (2020). Predictive Maintenance in the Industry 4.0: A Systematic Literature Review. *Computers & Industrial Engineering, 150*, 106889. https://doi.org/https://doi.org/10.1016/j.cie.2020.106889

3 Digital Image Processing Stages to Develop Augmented Reality Applications

3.1 INTRODUCTION

Nowadays, software tools such as "Adobe Aero" (Adobe, 2023), "Meta Spark Studio" (Meta, 2023), and "Aumentaty" (Aumentaty, 2023) are available in the market to implement AR applications quickly. However, creating an entirely usable AR application requires considerable effort regarding programming and computer vision skills.

Computer vision is a branch of artificial intelligence. A computer vision system allows the simulation of the human visual system using software and hardware. Therefore, computer vision aims to build artificial systems to extract information from digital images (Szeliski, 2010). In addition, computer vision exploits the techniques of digital image processing science to modify the quality of acquired images using mathematical algorithms (Gonzalez & Woods, 2006).

The reader should not confuse computer vision and digital image processing, because they are different. Digital image processing is used if the goal is to modify the pixels of an image. Computer vision is employed if the goal points to recognizing objects or detecting defects. Therefore, digital image processing is a subset of computer vision. This chapter explains the fundamental digital image processing stages for building an AR application.

3.2 THE STAGES TO DEVELOP AR APPLICATIONS

Figure 3.1 depicts the eight stages of developing an AR application, including: (i) camera calibration, (ii) image acquisition, (iii) preprocessing, (iv) segmentation, (v) representation and description, (vi) recognition, (vii) tracking, and (viii) registration.

Similar to the proposal by Chen and Lin (2019), the first six stages are conducted offline. As a result, a dictionary is built that includes the known markers and their associated virtual models. Then, stages two to eight are conducted online to create the AR experience.

3.2.1 CAMERA CALIBRATION

In an AR scene, a virtual object is superimposed into the real world to enhance or augment what is being seen by the user. The inserted virtual object must be at least

DOI: 10.1201/9781003435198-3

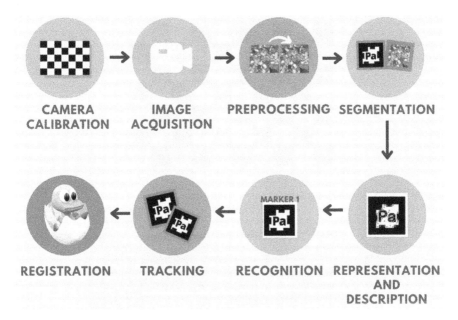

CAMERA   IMAGE   PREPROCESSING   SEGMENTATION
CALIBRATION   ACQUISITION

REGISTRATION   TRACKING   RECOGNITION   REPRESENTATION
    AND
    DESCRIPTION

FIGURE 3.1   Digital image processing stages to develop AR applications.

congruent with the scene regarding size, position, and orientation. Otherwise, the AR scene would not be meaningful. Virtual objects should naturally coexist with real objects so the user can feel satisfied with the AR experience. Camera calibration is the technique that helps to align and superimpose virtual objects accurately inside the real scene (de Oliveira et al., 2019).

The performance of a computer vision system depends mainly on the robustness of the camera calibration stage. Calibration is the process of creating a mathematical model that computes a camera's geometric and optical features, which are called the intrinsic (internal) parameters; the position and orientation of the camera regarding the world coordinate system are called the extrinsic (external) parameters and the distortion coefficients. In summary, calibration computes the pixel coordinate of a point inside the image and its correspondence to the real-world coordinate (Song et al., 2013).

The intrinsic parameters describe the camera operation, including the optical center, focal length, scale factor, skew, and lens distortion coefficients. On the other hand, extrinsic parameters determine the camera's orientation (rotation) and position (translation) and include the translation vector and the rotation matrix.

Calibration serves to remove lens distortion effects and obtain measures of the planar objects inside the digital images. It is essential to ensure that the computed parameters are quite similar to the real ones to ensure accurate measurements. The most common method to calibrate a camera in AR is employing a calibration pattern (marker). Estimating the camera model requires a set of points with three-dimensional coordinates and corresponding coordinates in pixels. The correspondences are obtained using a calibration pattern like the checkerboard shown in Figure 3.2. In

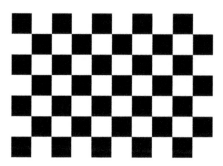

FIGURE 3.2   The calibration pattern.

FIGURE 3.3   Calibration pattern at different positions, orientations, and illumination.

the calibration pattern, precisely the number of vertices in the horizontal and vertical directions, and the spacing between the vertices must be known.

The calibration process begins by taking between 10 and 20 images of the calibration pattern at different positions, orientations, and illumination, as shown in Figure 3.3, where the size of the checkerboard square is 25 millimeters.

Then, each inner square corner on the checkerboard is detected, as shown in Figure 3.4. The images with incorrect detections are eliminated. The distorted images are corrected with the method proposed by Ahmed and Farag (2005). Afterward, the calibration process described by Zhang (2000) is performed. As a result, the calibration matrix k (intrinsic properties) shown in equation 3.1 and the rotation-translation matrix Rt (extrinsic properties) shown in equation 3.2 are obtained.

$$k = \begin{bmatrix} f_x & s & c_x \\ 0 & f_y & c_y \\ 0 & 0 & 1 \end{bmatrix} \qquad (3.1)$$

FIGURE 3.4 Detection of square inner corners in the calibration pattern.

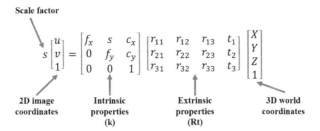

FIGURE 3.5 The process of mapping a 2D point to a 3D point.

where k is the calibration matrix, f_x and f_y are the focal length in x and y expressed in pixels, s is the skew parameter equal to zero, and (c_x, c_y) are coordinates of the optical center frequently localized in the center of the image.

$$Rt = \begin{bmatrix} r_{11} & r_{12} & r_{13} & t_1 \\ r_{21} & r_{22} & r_{23} & t_2 \\ r_{31} & r_{32} & r_{33} & t_3 \end{bmatrix} \tag{3.2}$$

where r's are the rotation coefficients that determine the camera orientation and t's are the rotation coefficients that determine the position of the camera projection center in the world coordinate system. In summary, the extrinsic matrix transforms the world coordinate system into the camera coordinate system, and this mapping is known as the camera pose. The extrinsic matrix transforms points from the camera coordinate system to the pixel coordinate system. Figure 3.5 depicts mapping a 2D point (u, v) to a 3D point (x, y, z).

In markerless AR, the calibration process employs the device sensors. The accelerometer, gyroscope, and compass calculate and track the device's orientation in real space.

3.2.2 Image Acquisition

The most important raw material for creating an AR application is the video stream acquired by a digital camera. The video is generated by reproducing image frames on a screen. Typically, the frames are reproduced at a speed of 30 frames per second (FPS). The frame rate defines the number of images a device can display in one second (Wandell et al., 2002). Since most current acquisition devices are digital cameras, each frame captured is a digital image in the red, green, and blue (RGB) space color.

The RGB color model comprises three primary colors to which the photoreceptor cones of the human vision system are most susceptible. A total of 3 bytes (24 bits) are employed to represent RGB color. One byte (8 bits) represents each of the three-color bands. Therefore, when performing the additive mix of the three colors, 16,777,216 colors can be represented (2^{24}). It is important to note that 16.7 million colors exceed considerably the 7 million colors a human can distinguish (Schwarz et al., 1987).

Because remembering millions of colors is complex, an alternative representation is employed. A color is defined by a triad of numbers between 0 and 255 corresponding to the 256 intensities each band can represent, where 0 is the darker color (black) and 255 is the lighter color (white). Therefore, the color is created by superimposing the intensity of one band on top of another (the colors must be summed up to obtain the newest one). Figure 3.6 shows an example of the color representation in RGB space.

A digital image is a bi-dimensional function $f(x, y)$ where x and y correspond to spatial coordinates of rows and columns, respectively. For example, the amplitude f in any coordinate pair (x, y) is known as the intensity or color. Because of the digitization process, all the values f can take are finite and discrete. Therefore, a digital image is formed from picture elements (pixels) that are spatially organized in a rectangular or square array (matrix).

A 3 × 3 RGB image is shown in Figure 3.7. As observed on the left, an image containing nine pixels with different colors is shown. In the middle, each RGB matrix is depicted. The color observed on the left is obtained by superimposing the R matrix on the G matrix and the G matrix on the B matrix. Because each band employs 8 bits, 3 bytes are needed to represent a color. The color represented with 24 bits is shown on the right-hand matrix. For example, the color red in coordinate (1, 1) is represented in binary as 1111 1111, 0000 0000, 0000 0000. Therefore, 16,711,680 is obtained when converting the binary number to decimal.

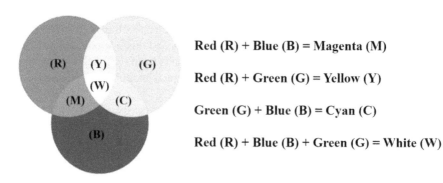

Red (R) + Blue (B) = Magenta (M)

Red (R) + Green (G) = Yellow (Y)

Green (G) + Blue (B) = Cyan (C)

Red (R) + Blue (B) + Green (G) = White (W)

FIGURE 3.6   RGB color space.

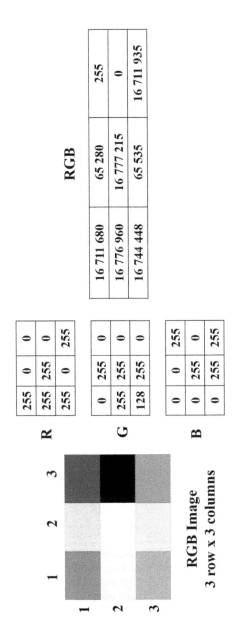

FIGURE 3.7 The representation of an RGB image.

It is essential to highlight that the image's origin is located on (1, 1) coordinate. Therefore, when a color is created, three planes must be defined. For example, for the color red, a 255 value is located in (1, 1, 1), a 0 in (1, 1, 2), and a 0 in (1, 1, 3). The last value of the triad defines the image color plane.

When the AR experience is activated, the camera of the technological device is switched on, and the image acquisition process starts. The acquisition process employs the digital camera's default resolution unless the user selects a different setting. Then, the video stream is delivered to the preprocessing stage.

3.2.3 PREPROCESSING

Preprocessing is the operation of touching the pixels' values inside an image to increase the probability of obtaining success in the following digital image processing stages (Gandhi et al., 2019). Frequently, if the color is not an important feature to distinguish an object inside the scene, the RGB is transformed into a grayscale image.

A grayscale image can represent 256 colors (2^8). Where 0 is black, and 255 is white. A grayscale image can be obtained from an RGB in two ways. The first method separates the RGB color bands, and the second employs equation 3.3.

$$0.299 * R + 0.587 * G + 0.114 * B \qquad (3.3)$$

Figure 3.8 depicts an RGB image and the corresponding grayscale images when the RGB bands are separated. The grayscale image obtained using equation 3.3 is shown on the upper right. The images obtained by separating the R, G, and B bands are shown at the bottom.

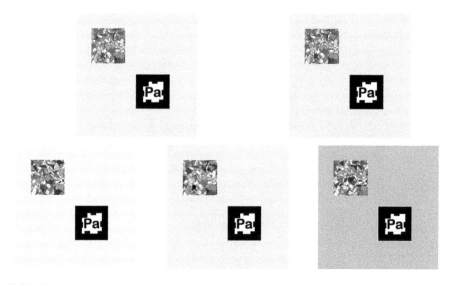

FIGURE 3.8 An RGB image and the grayscale images were obtained by applying equation 3.3 (top) and separating the R, G, and B bands (bottom).

The grayscale image can be converted to a binary image by defining a threshold. A binary image can contain only two colors (2^1), where 0 is black and 1 is the color white. The main problem with binarization is determining a threshold. The threshold serves as a condition to convert a grayscale intensity to zero or one. One of the most employed methods to automatically determine a threshold was proposed by Otsu (1979). The process proposes to review all values between 2 and 254 and select the one that minimizes the intra-class variance. The Otsu method can be applied globally or locally by defining image patches.

Half of the full-grayscale spectrum (128) is usually employed as a threshold. However, an effective threshold depends on the distribution of gray values. Therefore, all the values >= threshold are changed to white, and the remaining values are changed to black. Figure 3.9 shows the result of binarizing the grayscale image on the upper right of Figure 3.8. The conversion was obtained using 64, 128, and 192 thresholds, respectively.

Another image preprocessing technique is cleaning a noisy image, which is called denoising. An image can be naturally contaminated with noise during the acquisition, transmission, or compression (Fan et al., 2019). Therefore, the image must be cleaned. The denoising operation can be conducted using a filter. The filtering process is also known as convolution. Filtering is one of the most employed techniques in digital image processing. In AR, two-dimensional odd-size filters are used.

An image contaminated with artificial noise is shown in Figure 3.10. Salt and pepper (white and black points) noise was artificially added with a probability of 2%.

FIGURE 3.9 Images binarize using 64, 128, and 192 thresholds, respectively.

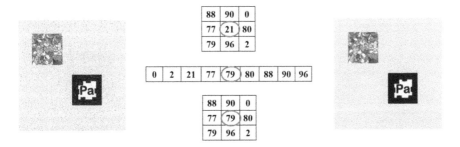

FIGURE 3.10 Image denoising with a median filter.

Denoising aims to clean up the image and preserve as many details as possible. The non-linear median filter was employed to clean the image.

As observed in the middle of Figure 3.10, a neighborhood of size 3 × 3 is selected starting from the image's upper left corner. The pixel in position (2, 2) is called the interest pixel, which will receive the response of the filtering process. The pixel values in the neighborhood are sorted from smallest to largest. Then, the median value is selected and assigned to the interest pixel. As a result, the high frequencies (noise) inside the image are attenuated. The process is repeated from left to right and upper to bottom until the median value changes all the pixels of the image. The result of denoising can be observed in the right part of Figure 3.10. Unfortunately, a blurred image is obtained due to median filtering. However, the artificial noise was almost eliminated.

Camera motion can also cause blurring in an image. Therefore, some parts of the image cannot be clearly observed. Restoring a blurred image into a cleaner (sharp) image employs a mathematical model named deblurring (Mahalakshmi & Shanthini, 2016). Figure 3.11 shows an example of blurred and deblurred images. Linear motion of the camera artificially created the deblurred image. The deblurred image was obtained using a Wiener Filter (Zhao et al., 2023).

Contrast is a property that can be enhanced in the image preprocessing stage. A dark image can be clarified (enhanced) by multiplying the pixels by a quantity that can increase its value. A clearer image can be obtained by multiplying all the pixel values by 1.5. On the other hand, an image can be darkened by multiplying the pixels by a quantity that can decrease its value. A dark image can be obtained by multiplying pixel values by 0.5.

Histogram equalization is a widely employed technique for contrast enhancement. In histogram equalization, the dynamic range of an image is altered to obtain a desired shape. A histogram is a representation of the distribution of the colors of an image. Histogram equalization spreads out the most frequent pixel values, so all pixel values are almost evenly distributed at the end. The right part of Figure 3.12 shows the result of equalizing the histogram on the left. Figure 3.13 shows the result of contrast enhancement using histogram equalization.

FIGURE 3.11   Examples of blurred and deblurred images.

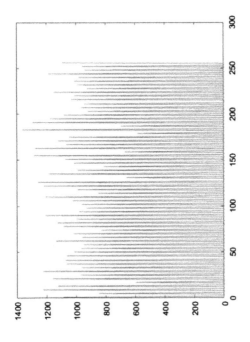

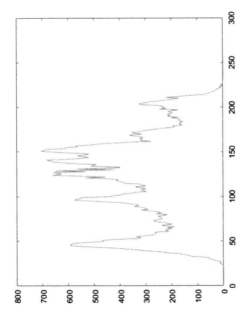

FIGURE 3.12 Image histogram equalization.

FIGURE 3.13   Image contrast enhancement.

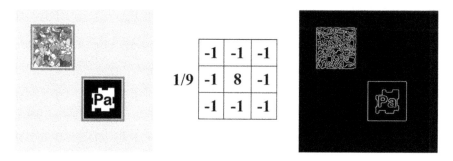

FIGURE 3.14   Segmentation stage using a high-pass filter.

The selection of the operations conducted in the preprocessing stage must be performed with great care to obtain acceptable results. The image obtained after this stage is the input to the segmentation stage.

3.2.4   SEGMENTATION

Segmentation is the process of separating all the objects inside the scene into multiple non-overlapping regions. The aim is to perform a labeling operation to determine the class to which each pixel pertains (Fu & Mui, 1981). The segmentation stage stops when all the objects inside a scene have been isolated. Commonly, segmentation is conducted considering the discontinuities (edges) or similarities (regions) of the image pixel values (Gonzalez & Woods, 2006).

Figure 3.14 shows an example of the output of the segmentation stage for two markers using edge detection. As observed on the left, ideally, the areas covered by the two markers inside the image must be detected, separated, and marked. A traditional high-pass filter was employed to obtain the resulting image of the edges on the right.

Figure 3.15 shows an example of region-based segmentation using connected component labeling, also called blob extraction. Blob extraction is based on graph theory, where the graph contains edges and vertices and serves to detect and count connected

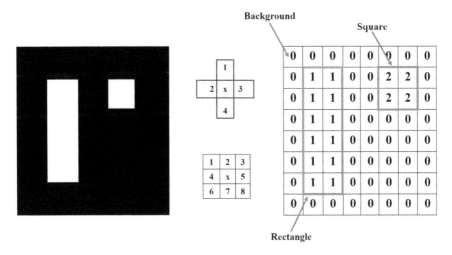

FIGURE 3.15 Segmentation using component-connected labeling.

regions (blobs) in binary images. The four or eight neighborhoods are employed to determine whether a component is connected or not (Samet & Tamminen, 1988).

The neighbors for each nonzero pixel are checked; at the end, each pixel has a label. However, this step is conducted at least one more time to avoid some connected regions having different labels. As a result, adjacent pixels have the same label, and the pixels from different regions have different labels (Shapiro, 1996). All the background was labeled with zero, the rectangle with one, and the square with two. Therefore, two objects were segmented inside the image.

From Figure 3.15, it is essential to highlight that the names of rectangle, square, and background were employed only to facilitate readers' understanding. In the segmentation stage only, the regions occupied by the objects are known, not the objects' names.

The novel techniques employ semantic segmentation to conduct the AR segmentation stage. Semantic segmentation is a deep learning algorithm that divides (labels) the image into pixels corresponding to pre-learned classes (Ko & Lee, 2020).

3.2.5 Representation and Description (Feature Extraction)

Each region obtained from segmentation should be described using features. The operation is conducted to extract the most relevant internal or external features essential to describe and distinguish one class of objects from another. Feature extraction is a pattern recognition task that can be considered a dimensionality reduction operation. At the end of this stage, the digital image can be discarded because the output is a vector with N features for each region (Kumar & Bhatia, 2014).

A feature is part of an object used to identify it. For example, a person can be described in terms of height, weight, skin color, and race. Therefore, a robust descriptor avoids the generation of different feature codes for objects in the same class

(Humeau, 2019). The main challenge is that the extracted features must be invariant to rotation, scale, translation, and contrast changes. Additionally, it is desired that the set of features extracted should be as small as possible. According to Tuytelaars and Mikolajczyk (2008), good features should have the properties of repeatability, distinctiveness, local, quantity, accuracy, and efficiency.

Color, texture, and shape are the common features extracted from images. Also, image corners are important features for object recognition and tracking because corners are invariant to rotation, translation, scale, illumination, and noise. A corner is the intersection of two edges. Moreover, corners are regions of the image with many variations in all directions. In image processing, a corner is also called an interest point (Bansal et al., 2021).

The method proposed by Harris is one of the most employed for corner detection and is an improvement of the Movarec proposal (Moravec, 1980). Corners can be detected inside an image by first detecting the edges. Then, a corner is detected in the place where two edges meet. Harris is based on the auto-correlation matrix and searches for the intensity changes produced in each pixel for a size-defined region. The input to the Harris corner detection is a grayscale image smoothed with a Gaussian filter. Then, the edges are extracted employing the Sobel or Canny algorithm. A 3×3 window is built around each image pixel to compute an intensity function. Finally, all the pixels that exceed a certain threshold and that are also the local maximum within the window must be found (Harris & Stephens, 1988).

Figure 3.16 shows the result of Harris corner detection on two different markers and its corresponding representation rotated 90 degrees counterclockwise.

FIGURE 3.16 Example images highlighting interest points obtained with Harris corner detection.

The location of each corner is stored in a database or feature vector. The scale-invariant feature transform (SIFT) and the speed up robust feature (SURF) are other feature extraction methods used in AR.

SIFT was developed by Lowe (1999) and extracts 128 features for each image interest point. The algorithm performs two main stages: (i) extraction of the interest points (key points) and (ii) description of the region around each key point. SIFT is invariant to rotation, scale, illumination, and viewpoint.

Key point extraction detects image regions where significant gradient changes occur on both sides of the point. Therefore, blobs and corners are detected at different scales employing a difference of Gaussians (DoG). Then, a local histogram of gradients (HoG) is calculated to describe the region around each key point. The dominant orientation of the gradient vectors in the neighborhood around the key point determines the invariance to rotation. Therefore, the orientation is employed to compute the histogram. The main drawback of SIFT is the great computational time employed to extract the features.

Figure 3.17 shows the features extracted using SIFT on the original marker, which is rotated 45 degrees counterclockwise. It is important to observe that many points located far away from the object were detected in the rotated image.

SURF was proposed by Bay et al. (2006) and is based on the Hessian matrix for scale and location. SURF is an improvement on SIFT because it is three times faster, but instead of using a Gaussian pyramid, it uses a Haar wavelet. SURF obtains 128 or 64 features from each key point, executing three main stages: (i) extraction of the key points, (ii) orientation assignment, and (iii) descriptor extraction. The key point extraction and scale determination are conducted with the determinant of the Hessian matrix. The orientation invariant point is calculated using the Haar wavelet on the x and y directions in a circular region with radio $6s$, where s is the keypoint scale. The dominant orientation is obtained by summing up all the results inside a sliding window that covers a $\pi/3$ angle. The descriptor extraction is conducted by building a square region centered on the key point with a size $20s$. The region is divided into

FIGURE 3.17   SIFT features extracted on two markers.

FIGURE 3.18 SURF features extracted on two markers.

four sub-regions. Then, the Haar wavelet is computed for x and y, and the results are smoothed using a Gaussian. The results for each sub-region are summed up, and the absolute value is calculated. In the end, each sub-region provides a vector v. The SURF descriptor is obtained through the union of the sub-region vectors.

Figure 3.18 shows the SURF features obtained on the original marker and the SURF features obtained with the marker translated 80 pixels and rotated 180 degrees counterclockwise.

In summary, feature extraction is an almost handmade process. Therefore, analyzing an object's most representative features can take a long time. Also, it is challenging to select which is the best feature detector to describe a particular object.

3.2.6 RECOGNITION

Recognition is the process of assigning a label (class or category) to an object inside an image. From the computational point of view, recognition is challenging because an object can appear inside the scene in different positions and contexts. Therefore, a machine learning algorithm is employed to recognize and correctly assign the object's label in less time (Weiss, 1993).

Computer recognition algorithms try to emulate the process conducted by humans for recognizing objects. First, the features of the object to be recognized are extracted. Then, the extracted characteristics are matched with a knowledge base to finally assign a label to the object. From the computational point of view, an image containing an unknown object should be compared or matched with the objects stored in a database. Then, the most similar is employed to assign a category to the object. Therefore, the features can be treated as points in a feature space, and the distance between the points can be employed to measure the similarity between two objects (Salari et al., 2022).

K-nearest neighbor (KNN) is a simple and efficient supervised pattern recognition algorithm that utilizes a proximity measure to determine the similarity between two samples. KNN comprises two stages: (i) training and (ii) testing. The training stage stores the information of each known sample with its label. The testing stage computes the distance of the new instance to every training sample to select the *K* closest results. *K* should be a positive integer major or equal to one. Finally, a majority rule among the *K* results is conducted to obtain the final decision (Zhang et al., 2018).

Two main decisions must be taken to implement KNN. The first is related to the similarity measure to be used. The Euclidean distance is frequently preferred, but the Mahalanobis or Minkowsky distances can also be employed. The second decision concerns selecting the *K* value, where an odd number is commonly recommended (Zhang et al., 2017).

Observe Figure 3.19 for an example of the KNN algorithm. Suppose two characteristics describing 10 cats and 10 dogs were extracted and stored in a database. Then, the characteristics are plotted. Cats are represented with red points, and dogs with green points. For the testing phase, three unknown samples should be recognized. The characteristics of the unknown samples are represented with blue points.

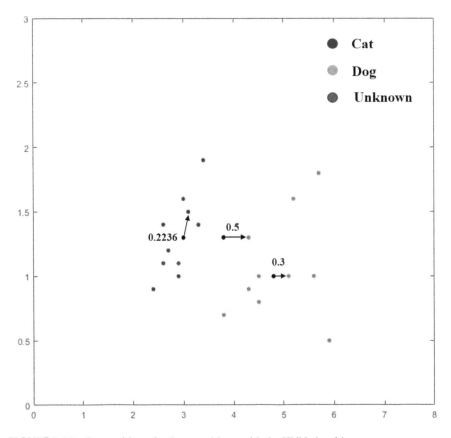

FIGURE 3.19   Recognition of unknown objects with the KNN algorithm.

For the example, the K value is equal to one. Hence, the Euclidean distance (D) from each unknown sample to all the known samples is computed using equation 3.4.

$$D\left(p_1, p_2\right) = \sqrt{\left(x_2 - x_1\right)^2 + \left(y_2 - y_1\right)^2} \qquad (3.4)$$

All the computed distances are ordered ascendant, and the k minor values are selected. As can be observed from the first unknown sample is located on (3, 1.3). Therefore, the $K = 1$ nearest neighbor is located on (3.1, 1.5) with 0.2236 distance and recognized as a cat. The second unknown sample located in (3.8, 1.3) is recognized as a dog because its nearest neighbor is (4.3, 1.3) with a 0.5 distance. The third unknown sample located in (4.8, 1) is recognized as a dog because its nearest neighbor is (5.1, 1) with a 0.3 distance.

The challenge of selecting the K value can be observed for unknown sample two. If $K = 3$ is selected, then the three minor distances obtained correspond to samples (4.3, 1.3), which is a dog, (3.3, 1.4), which is a cat, and (3.8, 0.7), which is a dog. Therefore, the voting strategy determines that dogs have two votes and cats have one vote. Then, it is confirmed that the sample is a dog.

Muja and Lowe (2009) observed that nearest-neighbor matching is a time-consuming component in high-dimensional spaces. Moreover, they observed that the linear search (brute force) was the faster algorithm to solve the problem. Therefore, an improvement of the KNN search for high dimensional spaces called the fast library for approximate nearest neighbors (FLANN) was proposed. FLANN is a heuristic approach that takes as input any data set and the necessary degree of precision to determine the best algorithm and optimum parameters.

The randomized kd-tree algorithm or the hierarchical k-means tree algorithm is the base of FLANN. Figure 3.20 shows an example of matching using FLANN with the 10 best correspondences. The key points were extracted using SURF, and FLANN

FIGURE 3.20 Feature matching with FLANN.

was employed as a matching descriptor. It is important to highlight that the matching is correctly conducted even when the second image is translated and rotated.

The stage of object recognition can be solved by employing artificial neural networks (ANN) (Abiodun et al., 2019) or support vector machines (SVM) (Pontil & Verri, 1998). Deep learning algorithms also can be used for the object recognition stage. Notably, a convolutional neural network (CNN) can be used for recognition purposes in AR (Dash et al., 2018; Estrada et al., 2022).

3.2.7 TRACKING

Once an object (marker) is segmented and recognized, tracking its movements across multiple video frames is imperative. Object tracking is a computer vision algorithm that tracks the movements of the recognized object in space or across different camera angles. A tracking algorithm must be fast, robust to occlusions, and able to recover when the object disappears, move fast, or move outside the edges of the video frames (Yilmaz et al., 2006).

For tracking purposes, an estimation of the past and current location of the object is computed. Then, the location is tracked while the object is visible. Object tracking can identify and follow multiple objects in an image. However, the complexity of the problem increases when more than one object needs to be tracked. The tracking process starts when the marker enters the scene and finalizes when the marker leaves the scene.

According to Wagner and Schmalstieg (2007), tracking fiducial markers is the most widely used technique in mobile AR applications. Marker-based tracking is less demanding than markerless-based tracking. For the case of marker-based AR, the marker is recognized employing simple techniques such as quadrilateral finding, and then the pose estimation using a homography is conducted. For the case of markerless AR, the key points extracted from SIFT, SURF, or other feature extractors are matched for recognition, then the outliers are removed. Finally, the pose estimation is conducted by computing a homography.

The Kanade-Lucas-Tomasi (KLT) algorithm can be employed for feature-based tracking (Lucas & Kanade, 1981; Tomasi & Kanade, 1991). First, Lucas and Kanade (1981) proposed a method for tracking an image patch. The aim is to conduct a local search using weighted gradients with an approximation to the second derivative of the image. As a result, the template image is aligned with the input image. Then, Tomasi and Kanade (1991) proposed a method to choose the best features for tracking. The features are selected if the eigenvalues of the gradient matrix are greater than a threshold.

The KLT algorithm comprises five steps: (i) find the corners where the eigenvalues are larger than a threshold, (ii) compute the displacement of each corner to the next frame with the Lukas-Kanade method, (iii) update the corner position and store the displacement of each corner, (iv) repeat steps (ii) and (iii), and (v) for each corner point return the long trajectories. Figure 3.21 shows the tracking of a marker located at different positions inside the digital image.

FIGURE 3.21   Marker tracking example.

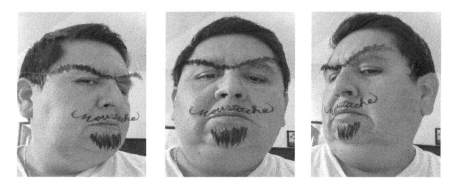

FIGURE 3.22   Example of AR face tracking.

Due to the accelerated use of social networks, face tracking has become a typical but challenging AR task. The aim is to employ high-precision algorithms to detect and track faces inside a video stream and then superimpose virtual content such as lenses, makeup, or a mask (Almeida et al., 2015). The algorithm first determines if a face is contained inside the image. Then, the location of the face is determined, and the facial features such as eyes, nose, and mouth are extracted. Finally, the virtual information is inserted according to the face pose. In the face tracking stage, knowing whose face it is is unimportant. That is, no recognition is performed; only detection is conducted.

Viola and Jones (2001) proposed an efficient and rapid face detection method based on the Haar cascade classifier. This method needs hundreds of positive (containing faces) and negative images (without faces) to train the classifier. Haar features (edges and lines) are extracted from each image. Therefore, each feature is obtained by subtracting the sum of pixels under a white rectangle from the sum of pixels under a black rectangle. Then, a boosting classifier is used to select the more relevant features. Finally, the result is optimized by connecting classifiers in a cascade to determine whether the features are part of an object (Huang & Cao, 2022).

Figure 3.22 shows an example of AR face tracking. As observed, the tracking allows inserting the virtual content in the exact location even when the face is in different positions inside the video stream.

3.2.8 REGISTRATION

Based on the tracking results, the virtual content should be displayed to the user in the final stage. The process of matching or aligning two or more images of the same scene taken from different viewpoints is known as registration (Brown, 1992). According to Azuma (1997), in an AR application, the virtual and real-world objects must be accurately aligned. If the virtual objects are not correctly scaled and orientated relative to physical objects in the real world, the illusion of AR will not be pleasant for the user.

According to Zitová and Flusser (2003), most registration methods comprise four steps: (i) feature extraction, (ii) feature matching, (iii) transform model estimation, and (iv) image resampling and transformation. The key points extracted in the feature extraction stage are prone to mismatches, causing the feature matching stage to be unsuccessful. Therefore, the RANdom SAmple Consensus (RANSAC) paradigm keeps only the inliers (erroneous data that lie in the interior of a statistical distribution) and rejects all outliers (data point that differs significantly from other observations) (Fischler & Bolles, 1981). Then, the homography is estimated from the remaining good matches. Finally, the AR scene is drawn (the virtual object is superimposed in the video stream).

The transformation that describes the relationship between two images is called homography. The homography matrix to compute a perspective transformation is shown in equation 3.5. The source and the destination images must be defined to compute a homography. Then, homography maps the points from the source image to the corresponding points in the destination image (Babbar & Bajaj, 2022).

$$H = \begin{bmatrix} h_{11} & h_{12} & h_{13} \\ h_{21} & h_{22} & h_{23} \\ h_{31} & h_{32} & h_{33} \end{bmatrix} \tag{3.5}$$

When applying the homography to the source image, it will get aligned to the destination image. Let us consider the set of key points (x_1, y_1) in the source image and the set of key points (x_2, y_2) in the destination image. The homography maps the key points in the source and destination image, as shown in equation 3.6.

$$\begin{bmatrix} x_1 \\ y_1 \\ 1 \end{bmatrix} = H \begin{bmatrix} x_2 \\ y_2 \\ 1 \end{bmatrix} = \begin{bmatrix} h_{11} & h_{12} & h_{13} \\ h_{21} & h_{22} & h_{23} \\ h_{31} & h_{32} & h_{33} \end{bmatrix} \begin{bmatrix} x_2 \\ y_2 \\ 1 \end{bmatrix} \tag{3.6}$$

As explained in subsection 3.2.2, the information collected from camera image acquisition can be contaminated with noise. Consequently, data cannot be well explained because they were affected by outliers, and one method to solve this problem is by fitting a line to the data points using RANSAC.

FIGURE 3.23   Registration between two markers using RANSAC.

All pairs of matching features are considered when the homography of two images is computed. However, some pairs correspond to valid matches (inliers), while others do not (outliers). The homography should be captured using only the inlier matches. RANSAC computes the homography for some iterations using a random sample of four correspondences. Each other correspondence is classified as an inlier or outlier depending on its concurrence with H. After all iterations, the iteration that contains the largest number of inliers is selected. Finally, H is recomputed using only the inliers. The main challenge is to decide when the correspondences are classified as inliers or outliers. Figure 3.23 shows the result of image registration between two markers using RANSAC.

After registration, the image acquisition stage starts, and the cycle is repeated while the video is captured.

REFERENCES

Abiodun, O., Jantan, A., Omolara, A., Dada, K., Umar, A., Linus, O., Arshad, H., Kazaure, A., Gana, U., & Kiru, M. (2019). Comprehensive Review of Artificial Neural Network Applications to Pattern Recognition. *IEEE Access*, 7, 158820–158846. https://doi.org/10.1109/ACCESS.2019.2945545

Adobe. (2023, February). *Adobe Aero*. www.adobe.com/products/aero.html

Ahmed, M., & Farag, A. (2005). Nonmetric Calibration of Camera Lens Distortion: Differential Methods and Robust Estimation. *IEEE Transactions on Image Processing*, *14*(8), 1215–1230. https://doi.org/10.1109/TIP.2005.846025

Almeida, D., Guedes, P., Silva, M., Silva, A., Lima, J., & Teichrieb, V. (2015). Interactive Makeup Tutorial Using Face Tracking and Augmented Reality on Mobile Devices. Proceedings of the XVII Symposium on Virtual and Augmented Reality, 220–226. https://doi.org/10.1109/SVR.2015.39

Aumentaty. (2023, February). *Aumentaty Solutions*. www.aumentaty.com/index.php

Azuma, R. (1997). A Survey of Augmented Reality. *Presence: Teleoperators and Virtual Environments*, *6*(4), 355–385. https://doi.org/10.1162/pres.1997.6.4.355

Babbar, G., & Bajaj, R. (2022). Homography Theories Used for Image Mapping: A Review. Proceedings of the 10th International Conference on Reliability, Infocom Technologies and Optimization (ICRITO) (Trends and Future Directions), 1–5. https://doi.org/10.1109/ICRITO56286.2022.9964762

Bansal, M., Kumar, M., Kumar, M., & Kumar, K. (2021). An Efficient Technique for Object Recognition Using Shi-Tomasi Corner Detection Algorithm. *Soft Computing*, *25*(6), 4423–4432. https://doi.org/10.1007/s00500-020-05453-y

Bay, H., Tuytelaars, T., & Van Gool, L. (2006). SURF: Speeded Up Robust Features. In Bischof, H. and Leonardis Aleš, P. A. (Eds.), *Computer Vision—ECCV 2006* (pp. 404–417). Springer Berlin Heidelberg.

Brown, L. (1992). A Survey of Image Registration Techniques. *ACM Computing Surveys*, *24*(4), 325–376. https://doi.org/10.1145/146370.146374

Chen, Y., & Lin, C. (2019). Virtual Object Replacement Based on Real Environments: Potential Application in Augmented Reality Systems. *Applied Sciences*, *9*(9), 1–21. https://doi.org/10.3390/app9091797

Dash, A., Behera, S., Dogra, D., & Roy, P. (2018). Designing of Marker-Based Augmented Reality Learning Environment for Kids Using Convolutional Neural Network Architecture. *Displays*, *55*, 46–54. https://doi.org/https://doi.org/10.1016/j.displa.2018.10.003

de Oliveira, M., Debarba, H., Lädermann, A., Chagué, S., & Charbonnier, C. (2019). A Hand-Eye Calibration Method for Augmented Reality Applied to Computer-Assisted Orthopedic Surgery. *International Journal of Medical Robotics and Computer Assisted Surgery*, *15*(2), e1969. https://doi.org/https://doi.org/10.1002/rcs.1969

Estrada, J., Paheding, S., Yang, X., & Niyaz, Q. (2022). Deep-Learning-Incorporated Augmented Reality Application for Engineering Lab Training. *Applied Sciences*, *12*(10), 1–19. https://doi.org/10.3390/app12105159

Fan, L., Zhang, F., Fan, H., & Zhang, C. (2019). Brief Review of Image Denoising Techniques. *Visual Computing for Industry, Biomedicine, and Art*, *2*(1), 7. https://doi.org/10.1186/s42492-019-0016-7

Fischler, M., & Bolles, R. (1981) Random sample consensus: A paradigm for model fitting with applications to image analysis and automated cartography. *Communications of the ACM*, *24*(6), 381–395.

Fu, K., & Mui, J. (1981). A Survey on Image Segmentation. *Pattern Recognition*, *13*(1), 3–16. https://doi.org/https://doi.org/10.1016/0031-3203(81)90028-5

Gandhi, M., Kamdar, J., & Shah, M. (2019). Preprocessing of Non-symmetrical Images for Edge Detection. *Augmented Human Research*, *5*(1), 10. https://doi.org/10.1007/s41133-019-0030-5

Gonzalez, R., & Woods, R. (2006). *Digital Image Processing* (3rd ed.). Prentice-Hall, Inc.

Harris, C., & Stephens, M. (1988). A Combined Corner and Edge Detector. *Proceedings of Alvey Vision Conference (AVC)*, pp. 147–152. https://doi.org/10.5244/C.2.23.

Huang, X., & Cao, X. (2022). Face Detection and Tracking Using Raspberry Pi based on Haar Cascade Classifier. Proceedings of the 37th Youth Academic Annual Conference of Chinese Association of Automation (YAC), 505–509. https://doi.org/10.1109/YAC57282.2022.10023612

Humeau, A. (2019). Texture Feature Extraction Methods: A Survey. *IEEE Access*, *7*, 8975–9000. https://doi.org/10.1109/ACCESS.2018.2890743

Ko, T., & Lee, S. (2020). Novel Method of Semantic Segmentation Applicable to Augmented Reality. *Sensors*, *20*(6), 1–18. https://doi.org/10.3390/s20061737

Kumar, G., & Bhatia, P. (2014). A Detailed Review of Feature Extraction in Image Processing Systems. Proceedings of the Fourth International Conference on Advanced Computing & Communication Technologies, 5–12. https://doi.org/10.1109/ACCT.2014.74

Lowe, D. (1999). Object Recognition from Local Scale-Invariant Features. *Proceedings of the Seventh IEEE International Conference on Computer Vision*, *2*, 1150–1157. https://doi.org/10.1109/ICCV.1999.790410

Lucas, B., & Kanade, T. (1981). An Iterative Image Registration Technique with an Application to Stereo Vision. *Proceedings of the 7th International Joint Conference on Artificial Intelligence (IJCAI)*, *2*, 674–679. https://hal.science/hal-03697340

Mahalakshmi, A., & Shanthini, B. (2016). A Survey on Image Deblurring. Proceedings of the International Conference on Computer Communication and Informatics (ICCCI), 1–5. https://doi.org/10.1109/ICCCI.2016.7479956

Meta. (2023, February). *Meta Spark Studio*. https://sparkar.facebook.com/ar-studio/.

Moravec, H. (1980). *Obstacle Avoidance and Navigation in the Real World by a Seeing Robot Rover* [B.Sc. dissertation]. Stanford University.

Muja, M., & Lowe, D. (2009). Fast Approximate Nearest Neighbors with Automatic Algorithm Configuration. Proceedings of the International Conference on Computer Vision Theory and Applications *(ICCVTA)*, pp. 1–10. https://api.semanticscholar.org/CorpusID:7317448

Otsu, N. (1979). A Threshold Selection Method from Gray-Level Histograms. *IEEE Transactions on Systems, Man, and Cybernetics*, *9*(1), 62–66. https://doi.org/10.1109/TSMC.1979.4310076

Pontil, M., & Verri, A. (1998). Support Vector Machines for 3D Object Recognition. *IEEE Transactions on Pattern Analysis and Machine Intelligence*, *20*(6), 637–646. https://doi.org/10.1109/34.683777

Salari, A., Djavadifar, A., Liu, X., & Najjaran, H. (2022). Object Recognition Datasets and Challenges: A Review. *Neurocomputing*, *495*, 129–152. https://doi.org/https://doi.org/10.1016/j.neucom.2022.01.022

Samet, H., & Tamminen, M. (1988). Efficient Component Labeling of Images of Arbitrary Dimension Represented by Linear Bintrees. *IEEE Transactions on Pattern Analysis and Machine Intelligence*, *10*(4), 579–586. https://doi.org/10.1109/34.3918

Schwarz, M., Cowan, W., & Beatty, J. (1987). An Experimental Comparison of RGB, YIQ, LAB, HSV, and Opponent Color Models. *ACM Transactions on Graphics*, *6*(2), 123–158. https://doi.org/10.1145/31336.31338

Shapiro, L. (1996). Connected Component Labeling and Adjacency Graph Construction. In Kong, T. Y., & Rosenfeld, A. (Eds.), *Topological Algorithms for Digital Image Processing* (Vol. 19, pp. 1–30). North-Holland. https://doi.org/https://doi.org/10.1016/S0923-0459(96)80011-5.

Song, L., Wu, W., Guo, J., & Li, X. (2013). Survey on Camera Calibration Technique. *Proceedings of the 5th International Conference on Intelligent Human-Machine Systems and Cybernetics*, *2*, 389–392. https://doi.org/10.1109/IHMSC.2013.240

Szeliski, R. (2010). *Computer Vision: Algorithms and Applications* (1st ed.). Springer-Verlag.

Tomasi, C., & Kanade, T. (1991). Detection and Tracking of Point. *International Journal of Computer Vision*, *9*(3), 137–154.

Tuytelaars, T., & Mikolajczyk, K. (2008). Local Invariant Feature Detectors: A Survey. *Foundations and Trends® in Computer Graphics and Vision*, *3*(3), 177–280. https://doi.org/10.1561/0600000017

Viola, P., & Jones, M. (2001). Rapid Object Detection Using a Boosted Cascade of Simple Features. *Proceedings of the 2001 IEEE Computer Society Conference on Computer Vision and Pattern Recognition (CVPR)*, *1*, I–I. https://doi.org/10.1109/CVPR.2001.990517

Wagner, D., & Schmalstieg, D. (2007). ARToolKitPlus for Pose Tracking on Mobile Devices. *Proceedings of the Computer Vision Winter Workshop (CVWW)*, 1–8.

Wandell, B., El Gamal, A., & Girod, B. (2002). Common Principles of Image Acquisition Systems and Biological Vision. *Proceedings of the IEEE*, *90*(1), 5–17. https://doi.org/10.1109/5.982401

Weiss, I. (1993). Geometric Invariants and Object Recognition. *International Journal of Computer Vision*, *10*(3), 207–231. https://doi.org/10.1007/BF01539536

Yilmaz, A., Javed, O., & Shah, M. (2006). Object Tracking: A Survey. *ACM Computing Surveys*, *38*(4), 13–es. https://doi.org/10.1145/1177352.1177355

Zhang, S., Li, X., Zong, M., Zhu, X., & Cheng, D. (2017). Learning k for KNN Classification. *ACM Transactions on Intelligent Systems and Technology*, *8*(3), 1–19. https://doi.org/10.1145/2990508

Zhang, S., Li, X., Zong, M., Zhu, X., & Wang, R. (2018). Efficient kNN Classification With Different Numbers of Nearest Neighbors. *IEEE Transactions on Neural Networks and Learning Systems*, *29*(5), 1774–1785. https://doi.org/10.1109/TNNLS.2017.2673241

Zhang, Z. (2000). A Flexible New Technique for Camera Calibration. *IEEE Transactions on Pattern Analysis and Machine Intelligence*, *22*(11), 1330–1334. https://doi.org/10.1109/34.888718

Zhao, S., Xing, Y., & Xu, H. (2023). WTransU-Net: Wiener Deconvolution Meets Multi-Scale Transformer-Based U-net for Image Deblurring. *Signal, Image and Video Processing*, *17*, 4265–4273 https://doi.org/10.1007/s11760-023-02659-z

Zitová, B., & Flusser, J. (2003). Image Registration Methods: A Survey. *Image and Vision Computing*, *21*(11), 977–1000. https://doi.org/https://doi.org/10.1016/S0262-8856(03)00137-9

4 Devices to Display Augmented Reality Experiences

4.1 INTRODUCTION

AR combines the real and virtual worlds, ideally making virtual models indistinguishable from the real ones. In order to reach its goal, AR should employ technological devices to display images that are close to reality and do not cause discomfort to the observer (Zhan et al., 2020). AR uses a set of optical, mechanical, and electronic components to generate images in the optical path between the observer's retina and the object to be augmented (Zhou et al., 2008). Therefore, displays are fundamental devices for designing AR applications. Generating AR through optical media must ensure almost normal vision of the environment. Consequently, the user perceives additional virtual elements through a digital display while the real world is simultaneously observed.

Researchers have tried implementing AR in displays based on optical systems, mobile devices, video, projectors, and, recently, the human eye's retina (Peillard et al., 2020). The process of selecting the ideal display is critical for the success of the AR application. Each display's characteristics, advantages, and disadvantages must be analyzed to conduct the selection. In addition, the field in which AR will be applied must also be considered. For example, considering the luminance requirement, a different display is employed indoors, outdoors, or both. Displaying short text could require less field of view (FOV) than a training application requiring wider FOV (Cakmakci & Rolland, 2006).

Determining the distance and position from the AR system to the human observer can also help to determine the display to be employed (Bimber & Raskar, 2005). For example, a retinal display could be used if the distance between the AR and the human must be short. On the other hand, a projector could be employed if the distance is long. Figure 4.1 shows the displays that can be employed considering the distance between users and the AR prototype.

The broadest way to categorize AR displays is as head-mounted and non-head-mounted. However, according to Carmigniani and Furht (2011) and Fang et al. (2023), AR visualization devices can be divided into head-mounted displays (HMDs), handheld displays (HDs), and spatial displays (SDs). Figure 4.2 shows a taxonomy for the display devices employed in AR. This chapter aims to inform readers about the different displays employed in AR experiences. Therefore, the technical details

DOI: 10.1201/9781003435198-4

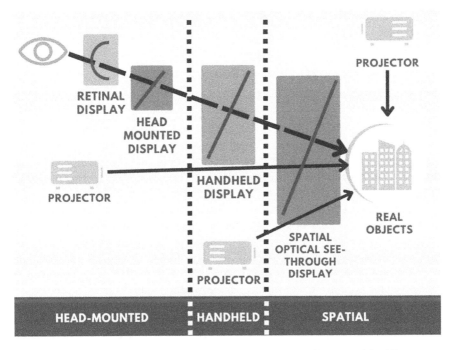

FIGURE 4.1   Types of displays considering the distance between the user and the AR prototype.

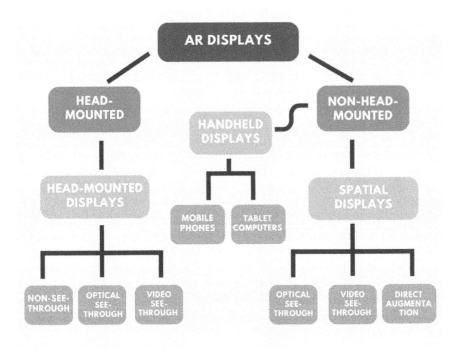

FIGURE 4.2   The taxonomy of the devices for displaying AR.

of each device are not addressed. Moreover, the following subsections explain each device type.

4.2 HEAD-MOUNTED DISPLAYS

An HMD is a wearable device with a display placed in front of one or both eyes integrated into eyeglasses or mounted on the user's head using a case similar to a helmet or a hat. Consequently, the digital content is superimposed on the user's FOV using computer vision algorithms. An optical system is employed to expand the FOV on the display, producing an imaginary screen that appears to be positioned several meters in front of the user. Hence, the user is involved in a hands-free immersive experience (Rolland & Hua, 2005).

In summary, an HMD must include a display, the optics for magnifying, and a combiner to mix the virtual objects with the physical world. The display units can be a cathode ray tube (CRT), a liquid crystal display (LCD), a micro LCD, liquid crystal on silicon (LCos), organic light-emitting diodes (OLED), transparent electroluminescent displays (TASEL), or laser source (Kress & Starner, 2013). The optics can be a reflective concave mirror, an eyepiece, or a projection lens system. The combiners can be a beam splitter, half-silver mirrors, semitransparent mirrors, dichroic mirrors, and holograms (Kress, 2019).

The design of an HMD is interdisciplinary and merges areas such as optical engineering, materials, manufacturing, user interface design, computer science, human perception, ergonomics, and physiology. Besides, the applications of HMDs span the fields of 3D scientific visualization, interactive control, education, entertainment, remote handling, telepresence, and training systems, among others (Cakmakci & Rolland, 2006).

The newest HMDs are gaining popularity because of the ease of moving anywhere, high-resolution display, lightweight, size, hands-free, and computational power. However, ensuring that these displays can be used for extended periods without producing eye strain, headaches, discomfort, reduced visual performance, or distractions is essential. Most current HMDs include the Android operating system and a camera to record the environment in which the user moves in real time. Moreover, all the users' movements can be tracked because the device is attached to the head. The user can control the device using a joystick, voice, or hand gestures. However, many research efforts had to be made for HMDs to have all these characteristics (Bauer & Rolland, 2023).

It all started in 1968 when Ivan Sutherland and his student Bob Strull designed what has been considered the first HMD. The Harvard laboratory project was named "The Sword of Damocles" and consisted of two tiny CRTs connected to a computer to generate graphics superimposed on the real scene using mirrors. A mechanical arm mounted on the ceiling held the HMD, allowing it to know the position and orientation of the user's head. Additionally, the device included a position sensor, three-dimensional graphics, stereoscopy, and the ability to navigate around an object and view it from various positions (Sutherland, 1968).

Just like a camera, an HDM has intrinsic and extrinsic parameters. Intrinsic parameters include focal length, aspect ratio, and optical center. On the other hand, extrinsic parameters include the position of the optic center and the orientation of the principal ray. HMDs also have an image plane, upon which pixels are drawn representing the rays of light from the scene striking the virtual film (Gilson et al., 2011).

Building an HMD similar to human visual ability is technologically challenging. Therefore, an HMD must be carefully designed for a given task, considering many parameters often conflicting. Also, the physiological factors of adaptation, convergence, binocular parallax, and monocular movement parallax and psychological factors related to retinal image size, linear perspective, aerial perspective, texture gradient, overlap, shapes, and shadows must be considered. Additionally, some technological issues that must be considered for the design are FOV, focal length, resolution, occlusion, and latency (Ens et al., 2014).

AR environments based on HMDs require stereoscopic vision and the ability to track the head movements and continuously update the display to reflect the user's movement through the environment. In addition, the user must be surrounded by visual stimuli with a suitable resolution, full color, appropriate brightness, and high-quality motion representation (Cheng et al., 2021).

Considering the user's perspective, also known as ocularity, HMDs can be monocular, biocular, or binocular. Ocularity measures the number of eyes needed to conduct a particular task. Therefore, the ocularity in HMDs is limited to two (Delabrida et al., 2016).

In monocular HMDs, the display is positioned in front of the right or left eye. Therefore, the image is viewed with one eye, and the other eye maintains a clear view of the environment. Monocular displays are frequently employed for information applications (Ward & Helton, 2022). In biocular HMDs, the display is positioned in front of both eyes. Therefore, the same image is viewed by both eyes. Biocular displays are recommended for proximity tasks (Kalich et al., 2006).

On the other hand, for binocular HMDs, the display is placed in front of both eyes. Hence, an independent image is displayed for each eye, generating a stereoscopic view. Binocular displays can be employed in almost all applications (Mon et al., 1993). Figure 4.3 provides a visual explanation of the three technologies.

In addition, HMDs can be divided into see-through and non-see-through devices. Moreover, two approaches can be used to generate and combine the real world with virtual objects with see-through devices. The first is the optical system, where beam splitters combine virtual objects with real scenes, and the second is the video system that employs a camera to capture the scene combined with virtual objects (Rahman et al., 2020). Figure 4.4 shows the three different types of HMDs.

4.2.1 Non-See-Through

Non-see-through HMDs (NST-HMDs), also known as occlusion or fully immersive displays, are mainly employed for entirely virtual reality experiences. These devices entirely block the users' view of the real world. Therefore, there is no visibility through the lenses, and they are bulky and heavy. An NST-HMD includes a stereoscopic display and a sensor for tracking the position and orientation of the user's

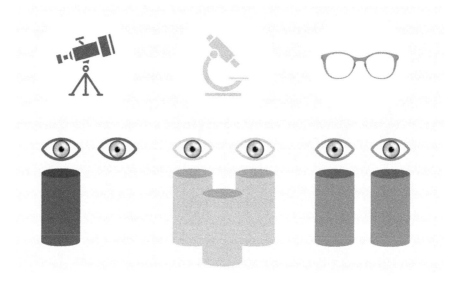

FIGURE 4.3 The different types of ocularity. Left: monocular; middle: biocular; right: binocular.

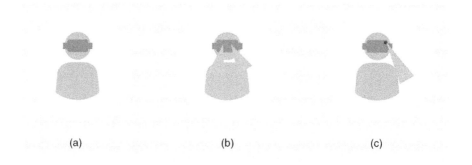

      (a)   (b)   (c)

FIGURE 4.4 Types of HMDs: (a) NST-HMD; (b) OST-HMD; (c) VST-HMD.

head. Moreover, the camera output mounted on the top is connected to a computer for processing. Also, the computer's video output is connected to the device, generating a whole immersive experience (Gilson et al., 2011).

Non-see-through HMDs are optically much simpler than other HMD types. Virtual object registration is unnecessary because the real world is invisible to the observer. However, these devices must be calibrated. Poor calibration can cause the distance to objects to be incorrectly estimated, causing premature fatigue or nausea (Rahman et al., 2020).

4.2.2 OPTICAL SEE-THROUGH

Optical see-through head-mounted displays (OST-HMDs) allow observing the world directly through semitransparent or transparent mirrors (combiners). Therefore, the virtual objects are simultaneously projected in front of the user's eyes (Grubert et al., 2018). The main components of an OST-HMD are the transmissive and reflective optical combiners. The transmissive property gives the user a direct view of the real world. The reflective property enables the user to observe the virtual superimposed objects. Therefore, the views of the physical world pass through the lens and graphically overlay the virtual contents to be naturally reflected in the user's eyes (Itoh et al., 2021).

Figure 4.5 shows a basic diagram of OST-HMD components (Doughty et al., 2022). As observed, optical combiners are located in front of the eyes. Hence, the user observes the real world through the combiners. The data from the user's head movements are computed with head trackers. The information from head locations is delivered to the scene generator. The scene generator processes the information and renders the virtual contents. Finally, the composed images pass to the monitor above the optical combiner to be projected to the user.

The first AR experience was created using an OST-HMD (Sutherland, 1968). Moreover, OST-HMDs were, for a long time, the most used devices to generate AR experiences (Zhou et al., 2008). The main advantage of using an OST-HMD is its simple structure. OST-HMDs do not require much computational power to present the image effectively. These devices can provide high-resolution images of the real world because the direct view of the physical world is maintained with minimum degradation. Moreover, OST-HMDs are exempt from power failures and offer no latency (delay) (Wang et al., 2023).

On the other hand, the main technological limitations of an OST-HMD are occlusion and light intensity. If a virtual object is displayed in front of a real object, it will appear to be a semitransparent ghost floating in front of the real object, causing

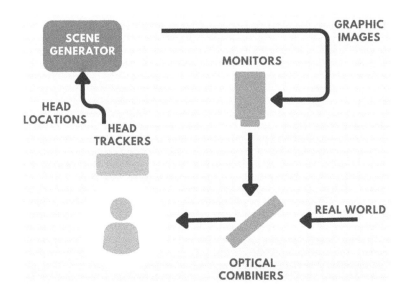

FIGURE 4.5 Conceptual diagram of an OST-HMD.

difficulty in seeing the virtual imagery. Moreover, if the virtual object is too bright compared to the ambient light, the user cannot see the real world clearly. Other severe limitations include updating of the screen, which does not respond appropriately to the movements of the head, lack of synchronization between what is seen through the glasses and the graphics, lack of resolve, the weight that prevents prolonged use, the need for frequent recalibrations, and limited FOV (Condino et al., 2020). The mentioned problems have been substantially reduced with advances in technology.

4.2.3 VIDEO SEE-THROUGH

In video-see-through HMD (VST-HMD), the vision of the world is captured with one or two miniature video cameras mounted on a helmet. Then, with computer vision techniques, the virtual models are electronically combined with real-world video representations and displayed on the screens located in the helmet. Hence, the user never directly sees the physical world because it is blocked and only sees the camera stream displayed through optics. Therefore, VST-HMD is known as fusion technology (Edwards et al., 1993).

Figure 4.6 shows a basic diagram of OST-HMD components (Azuma et al., 2001). As observed, the real-world scenes are acquired with digital cameras. The data from the user's head movements are computed with head trackers. Next, the information from head locations is delivered to the scene generator. The scene generator processes the information using computer vision algorithms and renders the virtual contents. Then, the videocompositor electronically mixes the real with the virtual content. Finally, the combined video is delivered to the monitor (nontransparent display) to be observed by the user.

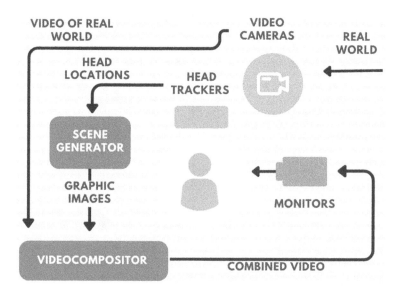

FIGURE 4.6 Conceptual diagram of a VST-HMD.

VST-HMD has many advantages, such offering a natural mixture of a real and a virtual scene because both are digital, and it can present a fully immersive perspective view with a wide viewing angle. When VST-HMD switches a see-through camera to another camera located in another place, the wearers experience moving to that location instantly to engage in remote operations (Pfeil et al., 2021). VST-HMDs can handle occlusion problems more easily than OST-HMDs and can provide high contrast and high fidelity without exhibiting the appearance of objects as if they were ghosts. Moreover, the depth of a scene can be calculated from multiple images.

Visual displacement and viewing deterioration are some drawbacks of VST-HMDs. Visual displacement occurs due to positional decoupling between the eyes and the camera. Viewing deterioration results from limitations in FOV, low display resolution, and display delay (Lee et al., 2023). Since the user is decoupled from reality, the social issue of turning off mutual eye contact has been experimented. Another disadvantage is that the camera's resolution limits the resolution of the captured image of the real world. Because VST-HMDs usually include two cameras, they are heavier than OST-HMDs. Moreover, people tend to underestimate distances when using VST-HMD.

4.3 HANDHELD DISPLAYS

Due to its weight and size, using an HMD extensively can cause fatigue and discomfort to the user. Therefore, handheld displays can be an alternative for experimenting with AR. Handheld devices are also known as mobile devices or digital assistants. Handheld devices are lightweight computational gadgets such as smartphones, tablets, or small projectors that users can hold and manipulate with their hands. Consequently, costly hardware to experiment with AR can be replaced with mobile devices (Sanna et al., 2015).

Handheld displays offer many advantages, such as portability, accessibility, sensors included (camera, gyroscope, accelerometer, GPS), and ease of manipulation (Wagner & Schmalstieg, 2003). Also, due to their low cost, almost everyone has a mobile device today. Indeed, young users prefer mobile devices because they can be used anytime, carried from place to place, and connected to the Internet all day.

The handheld devices' main limitation is usually the screen size, the computational power, and the battery's operating time. In addition, handheld devices limit user interaction because they need to point and hold the device with one hand and manipulate it with the other (Chatzopoulos et al., 2017).

The first mobile AR system was presented by Feiner et al. (1997). The AR process on mobile phones involves acquiring video, monitoring the computer application, and rendering and displaying images. Hence, VST techniques are employed to superimpose virtual content onto the real environment. According to Goh et al. (2019), mobile devices have become the primary output medium for AR.

4.3.1 MOBILE PHONES

Mobile phones, also known as smartphones or cellular phones, comprise a flat touchscreen interface operated by finger contact, physical buttons, and a digital camera. The text entry is conducted using an on-screen keyboard. Also, powerful processing, Internet connection, and sensors like GPS, compass, and accelerometer are included. A lithium-ion battery provides the power of a mobile device and includes a mobile operating system such as Android or iOS (Sarker & Wells, 2003).

The first mobile phone was manufactured by Motorola in 1973. The device was named Dynamic Adaptive Total Area Coverage (DynaTAC). DynaTAC was called "The Brick" because it weighed over a kilogram and measured 23 × 13 × 4.5 centimeters. Ten hours were required to recharge the device for a talk time of 30–35 minutes. The Motorola researcher Martin Cooper made the first mobile telephone call in human history (Harris & Cooper, 2019).

After over a decade of work, Motorola researcher Rudy Krolopp presented the first commercial mobile phone in 1983. The DynaTAC 8000x weighed 800 grams, measured 33 × 4.5 × 8.9 centimeters, and the sale price in 1984 was US$3,995. The battery offered one hour of talk time and 8 hours in standby mode. One year after its release on the market, 300,000 users had purchased the device. DynaTAC was succeeded by the much smaller MicroTAC Series in 1989 and later by the StarTAC Series (Murphy, 2013).

Currently, the mobile phone is considered one of the most important technologies for humanity because of two main features: talking and texting. Moreover, today, mobile devices are used more than fixed phones.

4.3.2 Tablet Computers

Tablets were historically considered an example of pen-computing technology. A tablet is a wireless portable personal computer with a touchscreen interface operated by a finger or stylus. A tablet uses a mobile operating system to perform its functions and includes a rechargeable lithium-ion battery. Modern tablets resemble smartphones, with the main differences being the weight and size of the screen. Moreover, tablets frequently do not support cellular network access (Atkinson, 2008).

The first conceptual design of a tablet called "Dynabook" was presented in 1972 by Alan Kay. The Dynabook was conceptualized with a thin body and a display manipulated with a stylus pen. Unfortunately, the technology to build the Dynabook did not exist then (Kay, 2011). The first tablet, called GRidPad, was launched on the market in 1989. The GRidPad weighed over 2 kilograms and measured 29.2 × 23.6 × 3.7 centimeters. The processor was a 20 MHz 386 and included a 10-inch screen with VGA resolution to display 32 gray levels. In addition, the GRidPad was powered by a battery with 3-hour autonomy and had a connector for a keyboard and floppy drive (Prey & Weaver, 2007).

In 1994, Fujitsu presented the Stylistic 500-tablet that includes an Intel processor and Windows 95 operating system. In 2002, Bill Gates introduced the Windows XP operating system for tablets. However, the history of tablet computers changed when, in 2010, Steve Jobs from Apple presented the iPad 1 as a multimedia platform to support books, movies, music, games, and web content. The iPad 1 weighed 1.5 pounds and was 0.5 inches thick. The LED screen was a brilliant 9.7 inches and included an A4 system on the chip (Gershon, 2013). The simplest model costs $499 US dollars and had no camera.

Also, in 2010, the first Samsung Galaxy Tab 7.0 was presented. The tablet included a 7-inch display and a 1GHz A8 Cortex processor. Also, a 2MP camera, a microSD card slot, and a USB 2.0 port were included (Karch, 2011). Samsung Galaxy Tab 7.0 cost $699.99 US dollars. Currently, Apple, Samsung, and Google are the leading tablet developers.

4.4 SPATIAL DISPLAYS

According to Raskar et al. (1998), spatial AR uses digital video projectors, holograms, optical elements, radio frequency tags, and tracking technologies to display graphical information in physical spaces such as walls or tables without requiring the user to wear or carry the display. Therefore, in spatial AR, most technology is separated from the system users and integrated into the environment using projectors or flat panel displays. Unlike HMDs and handheld displays, spatial displays do not support mobile applications.

Spatial display techniques are suitable for large presentations and exhibitions with limited interaction. Therefore, the high-resolution projection can span whole rooms with wide FOV (Bimber & Raskar, 2005). Moreover, many users can use a spatial display simultaneously to touch the virtual information physically; therefore, collaborative work is encouraged. Large buildings, cars, shoes, furniture, living creatures, and human dancers have been employed as surface projection targets.

Spatial AR, also called projection mapping, depends on geometrically and radiometrically aligning projectors and their surroundings to achieve a realistic, visually coherent augmentation (Bimber et al., 2008). Furthermore, in spatial AR, immersion is maximized because the information and virtual content are projected onto real objects using projected light. The immersion generated with spatial AR cannot be created with any other AR technology.

Spatial display systems are placed statically within an environment and, depending on how the environment is augmented, include video see-through, optical see-through, or direct augmentation (Marner et al., 2011).

4.4.1 VIDEO SEE-THROUGH SPATIAL DISPLAYS

VST spatial display systems employ conventional 2D and 3D monitors, televisions, or magic mirror metaphors to blend virtual content with real imagery for creating AR experiences. Screen-based AR is also called a window into the real world. The monitor size, the distance to the observer, and the spatial alignment relative to the observer limit the FOV in this AR experience. Therefore, it provides a low immersion degree (Tuceryan et al., 1995).

Screen-based AR uses a camera to capture the workspace, and then this information (video stream) is delivered to a computer for processing. The computer utilizes computer vision techniques to superimpose the virtual objects onto the video displayed on the monitor (Re et al., 2016).

A limitation of screen-based AR is the static nature of the display, and due to remote viewing, the user must create a mental mapping to understand the context and placement of virtually augmented information shown on the display when mapped in the real world (Bimber & Raskar, 2005).

The spatial display technique is now commonly applied in sports television as a complement to enhance the viewers' experience and help them understand the game rules. The first down line in American football, the glowing puck in ice hockey, the hack-eye line in tennis, the map of the track in a race, the review of the pitch in baseball, and displaying live statistics are examples of screen-based AR in sports (Chad, 2021).

The magic mirror is another example of VST spatial displays. In a magic mirror, the image of a person is acquired with a digital camera. Then, the image is displayed on a big screen enriched with virtual objects such as accessories or makeup. A magic mirror can create a virtual dressing room (Hashmi et al., 2020) or an application to explore anatomical structures in conjunction with medical images (Bork et al., 2017).

4.4.2   OPTICAL SEE-THROUGH SPATIAL DISPLAYS

OST spatial displays generate images aligned within the physical environment. Spatial optical combiners, such as planar or curved mirror beam splitters, transparent screens, or optical holograms, are essential to such displays (Bimber & Raskar, 2005).

A large half-silvered mirror beam splitter can be employed to generate a table-like display system aiming to extend the viewing and interaction space. Therefore, the user can view and interact with objects beyond a large optical combiner. Moreover, Johnson and Stewart II (1999) demonstrated that no statistically significant difference was observed in acquiring spatial knowledge between HMDs and OST spatial displays.

OST spatial displays offer advantages such as easier eye accommodation, high and scalable resolution, large FOV, prevented shadow casting, and improved ergonomics. On the other hand, the major drawback of OST spatial displays is the occlusion caused by the user's hands. Other disadvantages include optical distortion, calibration complexity, refraction, and virtual objects appearing to float around real objects (Bimber et al., 2001).

4.4.3   DIRECT AUGMENTATION

Direct augmentation or projection-based AR uses projection technology to augment and enhance 3D objects and spaces in the real world by projecting images onto visible surfaces. Direct augmentation uses sensors to track the user's movements and adjust the projection accordingly. The projected images can be computer-generated or photographic, either prerendered or generated in real time. Single, stereo, or multiple projectors can be used to increase the display area and enhance the image quality (Sand et al., 2016).

The main advantages of projection-based AR are that beautiful dynamic environments and shared experiences can be created. Moreover, because augmented objects have less relative motion, latency problems are significantly reduced. Other advantages include a theoretically unlimited field of view, a scalable resolution, and easier eye accommodation (Mine et al., 2012).

On the other hand, shadow-casting of the physical objects and of interacting users, restrictions of the display area, single focal plane located at a constant distance, increased complexity of consistent geometric alignment, and color calibration are the main disadvantages of projection devices (Bimber & Raskar, 2005). The dynamic range, frame rate, latency, spatial resolution, depth-of-field (DOF), and the device's displayable color gamut limit their applicability.

Direct augmentation encounters applications in industrial assembly, product visualization, medical training, museum display interactions, and architectural illustrations (Thomas, Billinghurst, & Haller, 2007).

REFERENCES

Atkinson, P. (2008). A Bitter Pill to Swallow: The Rise and Fall of the Tablet Computer. *Design Issues*, *24*(4), 3–25.

Azuma, R., Baillot, Y., Behringer, R., Feiner, S., Julier, S., & MacIntyre, B. (2001). Recent Advances in Augmented Reality. *IEEE Computer Graphics and Applications*, *21*(6), 34–47. https://doi.org/10.1109/38.963459

Bauer, A., & Rolland, J. (2023). The Optics of Augmented Reality Displays. In Nee, A. Ching, and Ong, S. (Eds.), *Springer Handbook of Augmented Reality* (pp. 187–209). Springer International Publishing. https://doi.org/10.1007/978-3-030-67822-7_8

Bimber, O., Encarnação, L., & Branco, P. (2001). The Extended Virtual Table: An Optical Extension for Table-Like Projection Systems. *Presence: Teleoperators and Virtual Environments*, *10*(6), 613–631. https://doi.org/10.1162/105474601753272862

Bimber, O., Iwai, D., Wetzstein, G., & Grundhöfer, A. (2008). The Visual Computing of Projector-Camera Systems. *Computer Graphics Forum*, *27*(8), 2219–2245. https://doi.org/10.1145/1401132.1401239

Bimber, O., & Raskar, R. (2005). *Spatial Augmented Reality: Merging Real and Virtual Worlds*. CRC Press.

Bork, F., Barmaki, R., Eck, U., Yu, K., Sandor, C., & Navab, N. (2017) Empirical Study of Non-Reversing Magic Mirrors for Augmented Reality Anatomy Learning. *Proceedings of the IEEE International Symposium on Mixed and Augmented Reality (ISMAR)*, pp. 169–176. 10.1109/ISMAR.2017.33

Cakmakci, O., & Rolland, J. (2006). Head-Worn Displays: A Review. *Journal of Display Technology*, *2*(3), 199–216. https://doi.org/10.1109/JDT.2006.879846

Carmigniani, J., & Furht, B. (2011). Augmented Reality: An Overview. In Furht, B. (Ed.), *Handbook of Augmented Reality* (pp. 3–46). Springer New York. https://doi.org/10.1007/978-1-4614-0064-6_1

Chad, G. (2021). *Augmented Reality in Sport Broadcasting* [Ph.D.]. Virginia Commonwealth University.

Chatzopoulos, D., Bermejo, C., Huang, Z., & Hui, P. (2017). Mobile Augmented Reality Survey: From Where We Are to Where We Go. *IEEE Access*, *5*, 6917–6950. https://doi.org/10.1109/ACCESS.2017.2698164

Cheng, D., Wang, Q., Liu, Y., Chen, H., Ni, D., Wang, X., Yao, C., Hou, Q., Hou, W., Luo, G., & Wang, Y. (2021). Design and Manufacture AR Head-Mounted Displays: A Review and Outlook. *Light: Advanced Manufacturing*, *2*(3), 350–369.

Condino, S., Carbone, M., Piazza, R., Ferrari, M., & Ferrari, V. (2020). Perceptual Limits of Optical See-Through Visors for Augmented Reality Guidance of Manual Tasks. *IEEE Transactions on Biomedical Engineering*, *67*(2), 411–419. https://doi.org/10.1109/TBME.2019.2914517

Delabrida, S., Loureiro, A., D'Angelo, T., Oliveira, R., Thomas, B., Carvalho, E., & Billinghurst, M. (2016). A Low Cost Optical See-Through HMD – Do-It-Yourself. Proceedings of the IEEE International Symposium on Mixed and Augmented Reality (ISMAR), 252–257. https://doi.org/10.1109/ISMAR-Adjunct.2016.0087

Doughty, M., Ghugre, N., & Wright, G. (2022). Augmenting Performance: A Systematic Review of Optical See-Through Head-Mounted Displays in Surgery. *Journal of Imaging*, *8*(7), 1–27. https://doi.org/10.3390/jimaging8070203

Edwards, E., Rolland, J., & Keller, K. (1993). Video See-Through Design for Merging of Real and Virtual Environments. Proceedings of IEEE Virtual Reality Annual International Symposium, 223–233. https://doi.org/10.1109/VRAIS.1993.380774

Ens, B., Finnegan, R., & Irani, P. (2014). The Personal Cockpit: A Spatial Interface for Effective Task Switching on Head-Worn Displays. Proceedings of the SIGCHI Conference on Human Factors in Computing Systems, 3171–3180. https://doi.org/10.1145/2556288.2557058

Fang, W., Chen, L., Zhang, T., Chen, C., Teng, Z., & Wang, L. (2023). Head-Mounted Display Augmented Reality in Manufacturing: A Systematic Review. *Robotics and Computer-Integrated Manufacturing*, *83*, 102567. https://doi.org/https://doi.org/10.1016/j.rcim.2023.102567

Feiner, S., MacIntyre, B., Höllerer, T., & Webster, A. (1997). A Touring Machine: Prototyping 3D Mobile Augmented Reality Systems for Exploring the Urban Environment. *Personal Technologies*, *1*(4), 208–217. https://doi.org/10.1007/BF01682023

Gershon, R. (2013). Digital Media Innovation and the Apple iPad: Three Perspectives on the Future of Computer Tablets and News Delivery. *Journal of Media Business Studies*, *10*(1), 41–61. https://doi.org/10.1080/16522354.2013.11073559

Gilson, S., Fitzgibbon, A., & Glennerster, A. (2011). An Automated Calibration Method for Non-See-Through Head Mounted Displays. *Journal of Neuroscience Methods*, *199*(2), 328–335. https://doi.org/https://doi.org/10.1016/j.jneumeth.2011.05.011

Goh, E., Sunar, M., & Ismail, A. (2019). 3D Object Manipulation Techniques in Handheld Mobile Augmented Reality Interface: A Review. *IEEE Access*, *7*, 40581–40601. https://doi.org/10.1109/ACCESS.2019.2906394

Grubert, J., Itoh, Y., Moser, K., & Swan, J. (2018). A Survey of Calibration Methods for Optical See-Through Head-Mounted Displays. *IEEE Transactions on Visualization and Computer Graphics*, *24*(9), 2649–2662. https://doi.org/10.1109/TVCG.2017.2754257

Harris, A., & Cooper, M. (2019). Mobile Phones: Impacts, Challenges, and Predictions. *Human Behavior and Emerging Technologies*, *1*(1), 15–17.

Hashmi, N., Irtaza, A., Ahmed, W., & Nida, N. (2020) An augmented reality based Virtual dressing room using Haarcascades Classifier. *Proceedings of the 14th International Conference on Open Source Systems and Technologies (ICOSST)*, pp. 1–6. 10.1109/ICOSST51357.2020.9333032

Itoh, Y., Langlotz, T., Sutton, J., & Plopski, A. (2021). Towards Indistinguishable Augmented Reality: A Survey on Optical See-through Head-Mounted Displays. *ACM Computing Surveys*, *54*(6), 1–36. https://doi.org/10.1145/3453157

Johnson, D., & Stewart II, J. (1999). Use of Virtual Environments for the Acquisition of Spatial Knowledge: Comparison Among Different Visual Displays. *Military Psychology*, *11*(2), 129–148. https://doi.org/10.1207/s15327876mp1102_1

Kalich, M., Lont, L., Bissette, G., & Jones, H. (2006). New Vision Performance Indicators of Biocular Head-Mounted Display Image Misalignment. In Brown, R., Marasco, P., Rash, C., & Reese, C. (Eds.), *Helmet- and Head-Mounted Displays XI: Technologies and Applications* (Vol. 6224, p. 622408). SPIE. https://doi.org/10.1117/12.665876

Karch, M. (2011). *Android Tablets Made Simple: For Motorola XOOM, Samsung Galaxy Tab, Asus, Toshiba and Other Tablets*. Springer.

Kay, A. (2011). A Personal Computer for Children of All Ages. Proceedings of the ACM Annual Conference, 1–11. https://doi.org/10.1145/800193.1971922

Kress, B. (2019). Optical Waveguide Combiners for AR Headsets: Features and Limitations. In Kress, B., & Schelkens, P. (Eds.), *Proceedings of Digital Optical Technologies 2019* (Vol. 11062, pp. 1–10). SPIE. https://doi.org/10.1117/12.2527680

Kress, B., & Starner, T. (2013). A Review of Head-Mounted Displays (HMD) Technologies and Applications for Consumer Electronics. In Kazemi, A., Kress, B., & Thibault, S. (Eds.), *Proceedings of Photonic Applications for Aerospace, Commercial, and Harsh Environments IV* (Vol. 8720, pp. 1–10). SPIE. https://doi.org/10.1117/12.2015654

Lee, J., Yeom, K., & Park, J. (2023). The Effect of Video See-Through HMD on Peripheral Visual Search Performance. *IEEE Access*, *11*, 85184–85190. https://doi.org/10.1109/ACCESS.2023.3304363

Marner, M., Smith, R., Porter, S., Broecker, M., Close, B., & Thomas, B. (2011). Large Scale Spatial Augmented Reality for Design and Prototyping. In Furht, B. (Ed.), *Handbook of Augmented Reality* (pp. 231–254). Springer New York. https://doi.org/10.1007/978-1-4614-0064-6_10

Mine, M., van Baar, J., Grundhofer, A., Rose, D., & Yang, B. (2012). Projection-Based Augmented Reality in Disney Theme Parks. *Computer*, *45*(7), 32–40. https://doi.org/10.1109/MC.2012.154

Mon, M., Warm, J., & Rushton, S. (1993). Binocular Vision in a Virtual World: Visual Deficits Following the Wearing of a Head-Mounted Display. *Ophthalmic and Physiological Optics*, *13*(4), 387–391.

Murphy, T. (2013). 40 Years After the First Cell Phone Call: Who Is Inventing Tomorrow's Future? *IEEE Consumer Electronics Magazine*, *2*(4), 44–46. https://doi.org/10.1109/MCE.2013.2273653

Peillard, E., Itoh, Y., Moreau, G., Normand, J., Lécuyer, A., & Argelaguet, F. (2020). Can Retinal Projection Displays Improve Spatial Perception in Augmented Reality? Proceedings of the IEEE International Symposium on Mixed and Augmented Reality (ISMAR), 80–89. https://doi.org/10.1109/ISMAR50242.2020.00028

Pfeil, K., Masnadi, S., Belga, J., Sera-Josef, J., & LaViola, J. (2021). Distance Perception with a Video See-Through Head-Mounted Display. Proceedings of the 2021 Conference on Human Factors in Computing Systems (CHI), 1–9. https://doi.org/10.1145/3411764.3445223

Prey, J., & Weaver, A. (2007). Tablet PC Technology–The Next Generation. *Computer*, *40*(9), 32–33. https://doi.org/10.1109/MC.2007.313

Rahman, R., Wood, M., Qian, L., Price, C., Johnson, A., & Osgood, G. (2020). Head-Mounted Display Use in Surgery: A Systematic Review. *Surgical Innovation*, *27*(1), 88–100. https://doi.org/10.1177/1553350619871787

Raskar, R., Welch, G., & Fuchs, H. (1998). Spatially Augmented Reality. *Proceedings of the First International Workshop on Augmented Reality*, 1–7.

Re, G., Oliver, J., & Bordegoni, M. (2016). Impact of Monitor-Based Augmented Reality for On-Site Industrial Manual Operations. *Cognition, Technology & Work*, *18*(2), 379–392. https://doi.org/10.1007/s10111-016-0365-3

Rolland, J., & Hua, H. (2005). Head-Mounted Display Systems. *Encyclopedia of Optical Engineering*, *2*, 1–14.

Sand, O., Büttner, S., Paelke, V., & Röcker, C. (2016). smARt.Assembly—Projection-Based Augmented Reality for Supporting Assembly Workers. In Lackey, S., & Shumaker, R. (Eds.), *Virtual, Augmented and Mixed Reality* (pp. 643–652). Springer International Publishing.

Sanna, A., Manuri, F., Lamberti, F., Paravati, G., & Pezzolla, P. (2015). Using Handheld Devices to Support Augmented Reality-Based Maintenance and Assembly Tasks. *Proceedings of the IEEE International Conference on Consumer Electronics (ICCE)*, 178–179. https://doi.org/10.1109/ICCE.2015.7066370

Sarker, S., & Wells, J. (2003). Understanding Mobile Handheld Device Use and Adoption. *Communications of the ACM*, *46*(12), 35–40. https://doi.org/10.1145/953460.953484

Sutherland, I. (1968). A Head-Mounted Three Dimensional Display. *Proceedings of the Fall Joint Computer Conference, Part* I, 757–764. https://doi.org/10.1145/1476589.1476686

Thomas, B., Billinghurst, M., & Haller, M. (2007). *Emerging Technologies of Augmented Reality: Interfaces and Design*. First edition, Hershey: Idea Group Publishing.

Tuceryan, M., Greer, D., Whitaker, R., Breen, D., Crampton, C., Rose, E., & Ahlers, K. (1995). Calibration Requirements and Procedures for a Monitor-Based Augmented Reality System. *IEEE Transactions on Visualization and Computer Graphics*, *1*(3), 255–273. https://doi.org/10.1109/2945.466720

Wagner, D., & Schmalstieg, D. (2003). First Steps Towards Handheld Augmented Reality. Proceedings of the Seventh IEEE International Symposium on Wearable Computers, 127–135. https://doi.org/10.1109/ISWC.2003.1241402

Wang, C., Hsiao, C., Tai, A., & Wang, M. (2023). Usability Evaluation of Augmented Reality Visualizations on an Optical See-Through Head-Mounted Display for Assisting Machine Operations. *Applied Ergonomics1*, *113*(1), 1–15. https://doi.org/https://doi.org/10.1016/j.apergo.2023.104112

Ward, M., & Helton, W. (2022). More or less? Improving Monocular Head Mounted Display Assisted Visual Search by Reducing Guidance Precision. *Applied Ergonomics*, *102*, 103720. https://doi.org/https://doi.org/10.1016/j.apergo.2022.103720

Zhan, T., Yin, K., Xiong, J., He, Z., & Wu, S. (2020). Augmented Reality and Virtual Reality Displays: Perspectives and Challenges. *IScience*, *23*(8), 1–13. https://doi.org/https://doi.org/10.1016/j.isci.2020.101397

Zhou, F., Duh, H., & Billinghurst, M. (2008). Trends in Augmented Reality Tracking, Interaction and Display: A Review of Ten Years of ISMAR. Proceedings of the 7th IEEE/ACM International Symposium on Mixed and Augmented Reality, 193–202. https://doi.org/10.1109/ISMAR.2008.4637362

5 Augmented Reality Development Cycle

5.1 INTRODUCTION

Requirements analysis is the first step in the development of any software product. However, it is usually challenging because clients are often unclear about how the software will work (Usman et al., 2020). Afterward, the system design is conducted using a macro-level scheme. Subsequently, the implementation begins with the writing of source code in the selected programming language. Verification is then performed to correct failures. Finally, the maintenance stage ensures the quality of the software (Khan et al., 2022).

Like any software, AR prototyping should follow a cycle. However, scientists have not yet agreed on the stages that should be included. An AR application should be characterized by quality, safety, and usability. Therefore, the application should be planned and designed carefully. The elements needed to implement an AR application can be broadly divided into two branches. The first is regarding the software elements, and the second is for the hardware elements (Marín et al., 2022).

Hardware elements refer to the device for acquiring the video stream and the display that will be employed to observe the augmented scenes. Software elements refer to the toolset used to build the AR prototype. This chapter addresses the software elements related to developing an AR prototype. Therefore, a development cycle is proposed. In the requirements analysis stage, the developer must define if what the client wants to solve can be addressed with RA. If the answer is yes, then the number of 3D models and their respective markers must be defined. Subsequently, the system must be programmed and implemented on Android or iOS. Finally, the prototype must be tested to verify its correct operation.

Figure 5.1 shows the proposed model that starts from the premise that the requirements analysis was carried out successfully. Therefore, the proposal begins with generating virtual models, followed by the markers design, programming, deployment, and user experience.

The following subsections present an explanation of how to conduct each stage. Moreover, the chapter ends by showing how to make a simple AR application with Meta Spark Studio.

DOI: 10.1201/9781003435198-5

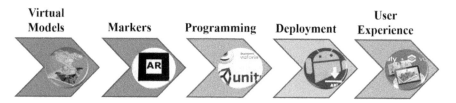

FIGURE 5.1   Proposal for the AR development cycle.

5.2 VIRTUAL MODELS

AR superimposes virtual models into the scene observed by the user. Therefore, virtual models play an essential role in AR and can be audio, video, texts, 2D drawings, or 3D models. The virtual models can be created using computer-aided design (CAD) software (Fiorentino et al., 2012).

CAD employs computer programs to create or modify virtual models of real-world products in 2D or 3D. Virtual objects can have dimensions, volumes, or weights but are purely digital representations (Imbert et al., 2013). There are two ways to create virtual models: (i) by hand employing modeling software and (ii) by using a 3D scanner.

The process of creating a virtual model starts with the conceptualization. The conceptualization is based on the ideas of the developer. Therefore, CAD software is employed to construct the virtual model by manipulating digital polygons. Then, the texture is applied to the virtual model to create realistic details. Also, lighting is employed to create the desired ambiance. When the virtual model is complete, rendering generates the final result.

In this stage, the software developer must define the number of virtual models the AR prototype will include. Moreover, the object type must be defined (text, video, 2D, or 3D) and whether it will contain animations. The goal is to create realistic objects, animated and texturized. The following subsections briefly describe the typical software for creating AR virtual models.

5.2.1 BLENDER

Blender is an open-source, cross-platform software used for modeling, lighting, rendering, and animating three-dimensional graphics. Blender was released in October 2002 under the General Public License (GNU GPLv2) terms. Blender's development continues daily thanks to a team of volunteers from around the world led by its inventor, Ton Roosendaal (Van Gumster, 2020).

Blender has its own game engine and all the tools for modeling and animating characters and creating background scenes. It has a video editor, integrates node editing technology, offers free plugins, and allows importing multiple 3D formats (Blain, 2019). Blender can be downloaded at www.blender.org.

At the time of writing this book, the newest version of Blender was 3.6.2. Blender offers color and textures to enhance models and scenes, aiming to produce realistic effects. Also, tools for 2D animation are included. The Phyton API is available for scripting and customization. However, like all software to generate computer graphics, Blender includes many buttons, settings, and working methods. Therefore, people consider it a software in which it is difficult to become an expert.

5.2.2 Autodesk Maya

Maya is a cross-platform professional software created by Autodesk that offers tools for developing 3D graphics, rendering, special effects, and animations. Maya was used to create scenes that have appeared in movies, television, and games. Most Oscar-nominated movies used Maya to create visual effects (Murdock, 2022).

The most important feature of Maya is how open it is to third-party software. Maya's power, expansion, and customization are other characteristics of its interface. However, since Maya is a professional software, it has the characteristic of being expensive. The trial version of Maya can be downloaded at www.autodesk.com/produ cts/maya/overview.

Maya Embedded Language (MEL) is employed for scripting and package personalization. Maya interface was created with Qt libraries (King, 2019). At the time of writing this book, the newest version of Maya was 2023.3.

5.2.3 SketchUp

Unlike other graphics and animation software, SketchUp is very user-friendly. SketchUp is a face-based 3D modeling and graphic design program. The multiplatform software was created in 1999 by the "Last Software" company. The first name was Google SketchUp, but in 2012, "Trimble Inc." acquired the program, and the name was changed to SketchUp (Chopra, 2007).

The main feature of SketchUp is being able to make 3D designs easily. The software offers a step-by-step tutorial to model and design the first virtual environment. In addition, it includes a gallery of objects, textures, and images ready to download (Chopra, 2012). The homepage for SketchUp is www.sketchup.com/.

SketchUp was written in the object-oriented programming language called Ruby. Therefore, with that language, users can write custom plugins. It can also be used for product design. The interface can be configured to suit each person's style and working style. Also, compared to other 3D design software, SketchUp is reasonably priced. At the time of writing this book, the newest version of SketchUp was 2023.0.

5.2.4 Cinema 4D

Cinema 4D is a multiplatform computer 3D animation, modeling, simulation, and rendering software developed by Maxon. The set of tools it offers is fast, powerful, and flexible, which makes the workflow accessible and efficient. Professional artists employ the application to create impressive 3D scenes (Szabo, 2012). The price of Cinema 4D is not as high as other software. The home page of the software is www. maxon.net/es/cinema-4d.

Cinema 4D includes MoGraph to create complex and abstract animations quickly and easily. For ease of use, Maxon offers tutorials and hundreds of quick tips. In addition, Cinema 4D is available in 11 languages. The software facilitates object cloning and generating movements without difficulty (McQuilkin & Powers, 2012). At the time of writing this book, the newest version of Cinema 4D was 2023.2.2.

The main virtues of Cinema 4D are modularity, high rendering speed, a highly customizable and flexible interface, and a vertical learning curve. Cinema 4D has a physics engine that allows for complex collisions and interactions between objects in a simple way.

5.3   MARKERS

Computer vision techniques are employed in AR to detect and recognize landmarks. Landmark recognition allows the integration of virtual content on the video stream observed by the user. In marker-based AR, the easiest way to recognize environmental landmarks is by using images designed for this purpose, called fiducial markers. However, in markerless-based AR, any object inside the environment can be employed as an AR trigger, which makes the process more complex (Pooja et al., 2020).

Fiducial markers must be designed carefully to facilitate detection and recognition stages. Therefore, bitonal markers are usually templates with a black square, a four-fold smaller white square in its center, and a simple image inside. A white edge is typically located outside the black square. The markers should be distinct enough not to be confused with the environment (Fiala, 2010). A marker must be printed on a white sheet or displayed in any way a camera can see.

In AR, markers must be detected within a large field of view regardless of whether they appear occluded or distorted. Therefore, the information included inside the marker must not be too dense. Strictly, any image editing software can be used to generate the markers. However, some applications help the generation of markers. Two examples are provided in the following subsections.

5.3.1   ARToolKit Marker Generator

ARToolKit recognizes and tracks the square markers in the video stream. Therefore, the ARToolKit marker generator is a tool for creating markers (Fiala, 2010). First, the user must select to create a pictorial or 2D barcode marker. The marker size in millimeters must be defined. Then, the marker resolution in dots per inch (DPI) is defined. The values can be 72 dpi or 300 dpi. After selecting the dpi, the marker pixel size is computed.

The user must define whether a border size is included. A percentage defines the size. The user must choose the color of the borders, which can be black or white. The user must define the barcode size if the barcode was selected. Then, the error checking and correction time (Hamming or parity) is defined.

Finally, the user must define whether single or multiple markers will be designed. The markers designed can be downloaded to the computer's hard disk. The tool can be accessed at https://au.gmented.com/app/marker/marker.php. Processing science to modify images uses mathematical algorithms.

5.3.2 Brosvision Marker Generator

Brosvision offers a tool to generate markers optimized for AR use. Each marker is generated randomly. Consequently, each marker is unique and complies with the recommendations for accurate and reliable tracking results. Markers can be generated in RGB or grayscale. Triangles, lines, and quadrangles are the base to generate markers. The designer must select whether the primitives are included in low, medium, or high quantities (Brosvision s.r.o., 2023).

The markers generated can be downloaded and used for free in AR applications. Brosvision Marker Generator can be accessed at https://brosvision.com/ar-marker-generator/.

5.4 PROGRAMMING

One of the most challenging decisions when building an AR prototype is selecting the software developer kit (SDK) on which to implement it. An analysis must be carried out to select an SDK that can fulfill the application's specific purpose. What is a fact is that the software to be used must include computer vision techniques to ensure that the virtual models are accurately inserted into the video flow observed through a screen.

SDKs offer the coding environment to implement all the core functionalities of the AR application. Numerous AR SDKs are available in the market, each with many different functionalities. However, the developer should consider the license type, the supported platform, the devices supported, and hardware requirements (Mladenov et al., 2018). In the following subsections, we briefly describe the features of eight platforms aiming to help programmers in their choice decisions.

5.4.1 Vuforia

Vuforia is perhaps the most popular SDK for developing AR applications because it supports most telephones, tablets, and HMDs. Vuforia employs robust computer vision algorithms for recognizing and tracking AR markers and computations for object superimposition in real time (Simonetti & Paredes, 2013). Qualcomm formerly introduced Vuforia in September 2014. However, PTC acquired Vuforia in October 2015.

The C++ QCAr library is the platform's core, containing targets and image rendering features. Vuforia has a free version (limited functionality) with which simple AR implementations can be made. However, it is best to have a payment plan contracted to use all the features.

Vuforia applications comprise the following elements: (i) camera, (ii) database, (iii) image converter, (iv) tracker, (v) renderer, (vi) application code, and (vii) target resources. The languages supported for development include C++, Java, Objective

C++, and .Net. Vuforia supports native Unity editor development to create Android or iOS applications (Hameed et al., 2022). At the time of writing this book, the newest version of Vuforia Engine was 10.17. Moreover, the trial version can be downloaded at https://developer.vuforia.com/.

5.4.2 WIKITUDE

Wikitude is a cross-platform SDK for creating AR applications launched in 2008 by an Austrian company. Initially, Wikitude focused on providing geolocation-based AR experiences using the Wikitude World Browser. However, in 2012, the Wikitude SDK was presented, including geolocation technologies and object recognition and tracking.

Developers can create customized solutions and perform the rendering employing third-party software. Wikitude was voted the best AR browser for four consecutive years (Madden, 2011). Wikitude employs the mobile device's display to visualize information that can be organized in worlds. The information comes from Wikipedia, X, or travel sites (TripAdvisor). The Wikitude 3D encoder encodes the 3D source file to a compressed binary format (.wt3) for describing 3D content. The format is optimized for fast loading and handling 3D content on mobile devices.

The trial version of Wikitude can be downloaded at www.wikitude.com/download/. At the time of writing this book, the newest version of Wikitude was 1.7.

5.4.3 ARKIT

ARKit is a free AR platform developed by Apple in 2017. ARKit uses cameras (front and rear) and device sensors to measure the environment's dimensions. Also, the platform can integrate virtual objects into real scenes by considering lighting conditions. Therefore, ARKit creates an experience as close to reality as possible (Permozer & Orehovački, 2019).

Developers can create applications for iPhone and iPad, and by employing the multiuser strategy, two devices can share virtual items. Visual Inertial Odometry (VIO) is the base technology ARKit uses to make the correspondences between real and virtual worlds. VIO mixes information from the motion sensing hardware with computer vision techniques (Wang, 2018).

Computer vision algorithms recognize the features (markers) inside the scene, track the features across video frames, and make comparisons with data obtained from motion sensing. As a result, a precise model of the device's position and motion is obtained. ARKit also allows the creation of geolocation-based applications.

ARKit can be downloaded at https://developer.apple.com/augmented-reality/arkit/. At the time of writing this book, the newest version of ARKit was Wikitude 6.

5.4.4 ARCORE

ARCore is a multiplatform SDK created by Google in 2018 for building AR applications. ARCore is a platform mainly for Android devices, including motion capture, environment estimation, depth comprehension, and light perception. ARCore can be used in popular development environments such as Unity, Unreal, Android, and iOS (Lanham, 2018).

ARCore uses the mobile device's camera to extract key points called features. The information from features is used to track movement over time. With the combination of tracking and inertial sensors, ARKit accurately computes the location where virtual objects must be inserted in the real scene (Linowes & Babilinski, 2017).

One of the crucial points to develop with ARCOre is to identify whether the mobile device where the implementation will take place is compatible. The list of mobiles that Google constantly updates must be consulted to conduct the verification.

ARCore can be downloaded at https://developers.google.com/ar?hl=es-419. At the time of writing this book, the newest version of ARCore was 1.38.

5.4.5 Aumentaty

Aumentaty is a free platform launched in 2012 by LAbHuman for creating and sharing AR content for Android and iOS. The main advantage of this platform is that anyone, whether they know how to code or not, can create an AR experience by following simple steps. Aumentaty comprises two software tools: (i) Aumentaty Creator and (ii) Aumentaty Scope (Aumentaty, 2023).

Aumentaty Creator is the tool to create scenes in which virtual objects are associated with markers. Using Creator, developers can design marker-based AR or geolocation-based AR applications. Moreover, the virtual objects can be images, texts, videos, or 3D models. Creator can be downloaded at www.aumentaty.com/community/es/. At the time of writing this book, the newest version of Creator was 1.3.6.

Aumentaty Scope is the viewer of the contents generated by Creator. Scope must be downloaded to the mobile device and is supported for Android and iOS. When Scope is executed in the mobile device, the user must find the experience created and then open it to experiment with AR.

5.4.6 Meta Spark Studio

Formerly named Spark AR in 2017, Meta Spark Studio is an SDK created by Meta to design AR effects and filters. The experience created can be shared on social networks such as Meta and Instagram. Meta offers templates that inexperienced programmers can use to generate an AR application rapidly.

The three main effects that can be designed with Meta Spark are people, world, and group for video calling. People effects use the device's front camera and respond to the movement of someone, including body, hand, and face tracker, segmentation, and deformation. World effects allow inserting virtual objects in the real video stream using the mobile device's back camera. The plane tracker of the target tracker must be employed to generate a world effect. Group effects are used for multipeer video calls on Instagram (Afshar, 2023).

Meta Spark Studio can be downloaded at https://spark.meta.com/. At the time of writing this book, the newest version of Meta Spark Studio was v162.

5.4.7 ADOBE AERO

Aero is an all-in-one free platform introduced by Adobe Inc. in November 2019 for creating and publishing AR experiences. Aero is available for iOS and in a beta desktop version for Windows and macOS through Creative Cloud. Aero offers 2D, 3D, and audio starter assets (Adobe, 2023).

Also, Aero allows importing files from Photoshop, Illustrator, or Substance in a transparent way. Programming skills are not needed to design an AR prototype. The platform uses the block-building approach to add interactivity to the immersive AR scenes.

The contents created can be shared with a link and experimented on any iOS device using Aero Clip. Moreover, Aero files can be exported to Xcode to extend the capabilities of the experience. The desktop application of Aero can be downloaded at www.adobe.com/mx/products/aero.html. At the time of writing this book, the newest version of the desktop Adobe Aero was 0.20.2, while for mobile devices, it was 2.23.6.

5.4.8 ARTOOLKIT

ARToolKit is a C/C++ open source library developed in 1999 by the University of Washington's Human Interface Technology Laboratory (HITLab) to develop AR applications (Kato & Billinghurst, 1999). ARToolKit was known as the most popular library for AR development and is the base for many AR platforms today.

ARToolKit employs computer vision techniques to conduct video tracking. Therefore, the camera's position and the orientation relative to the position of the markers can be calculated in real time. This way, the 3D models can be superimposed precisely onto the marker. Unfortunately, tracking can only be done for simple figures (Ximo, 2012).

ARToolKit can run on Windows, Linux, and Mac OS. Currently, ARToolKit is maintained as an open source project by ARToolWorks. The last version, 2.72.1, was released in 2008 and can be downloaded at www.artoolworks.com/index.html.

Due to ARToolKit's impact on AR, variants have been developed for different platforms and languages, such as FLARToolkit for Flash scripting, JSARToolkit for JavaScript., SLARToolkit for Silverlight, and NyARToolkit for Java, C# and Android.

5.5 DEPLOYMENT

Deployment is one of the most critical aspects of the software development process. Software deployment consists of delivering software from development to production environments, making the software available to end-users (Dearle, 2007). Deploying an AR application means making it accessible and ready for use on a target device through a public or private distribution channel. Deployment verifies whether the AR app can be installed correctly on the device to avoid compatibility issues (Javornik et al., 2021).

AR developers write the code to meet customer expectations, and before releasing the application, they conduct tests to ensure the application is optimized for running on mobile devices. Then, the application is packaged in a suitable format to be shared in an app store. Mainly, there are two alternatives to deploy on mobile devices: Android OS and iOS (Tracy, 2012).

5.5.1 ANDROID OPERATING SYSTEM

Android is an operating system for mobile devices developed by Android Inc. in 2004 and bought by Google in 2005. In 2007, Google announced the founding of the Open Handset Alliance (OHA). OHA is a consortium of technology and mobile devices companies to promote Android as a free open source operating system with support for third-party applications (Krajci & Cummings, 2013).

The main source code for Android is known as the Android Open Source Project (AOSP). Android contains a Linux-based kernel, middleware, libraries, an application programming interface (API) written in C, and a Java virtual machine called Dalvik. Android 1.1, called Petit Four, appeared in February 2009, and since then, Android versions have been named in alphabetical order and with the name of a dessert (Ableson et al., 2011). Android is the most used mobile operating system in the world.

The Android Application Package (APK) is needed to distribute apps run on devices with Android OS. The APK file is generated by compiling and packaging the source code in the container file. Therefore, an APK contains programs, certificates, resources, classes, assets, and manifest. APK files can be downloaded directly to the mobile device or from the Google Play Store (Deitel & Deitel, 2016).

A developer account with the Google Play Store is needed to deploy a mobile application on Android. Metadata for each application must be provided, including name, description, category, and icon. Applications uploaded to the store are not reviewed; hence, they are available immediately. After downloading the APK, the user only needs to open the file to start the installation and launching process. A security alert would be triggered if the file was not downloaded from the store, indicating that permission to install unknown apps must be activated. This can be solved by entering the device settings.

5.5.2 IPHONE OPERATING SYSTEM

The iPhone Operating System (iOS) was developed by Apple in 2007 for running applications natively on iPad and iPhone devices. The Apple iOS platform is based on a proprietary model known as a closed system because it can only be used on Apple devices. The iOS platform consists of four layers aiming for applications to communicate with hardware devices: (i) Core OS/Kernel, (ii) Core Services, (iii) Media Support, and (iv) the Cocoa Touch Interface layer (Bollapragada et al., 2000).

The iOS kernel is the XNU kernel of Darwin. iOS SDK contains the code, information, and tools needed to develop, test, run, and debug iOS apps. iOS 1.0 was presented with the first iPhone and did not allow app downloads. Currently, iOS is the second most used mobile operating system after Android (Feiler, 2014).

ARM architecture is the hardware platform for iOS. The iOS programmer needs to set up a MAC to write iOS apps. An iOS application can be programmed with Swift,

Objective-C, C#, Java, or JavaScript. Moreover, Xcode provides a basic editing, compilation, code debugging environment, and a simulator that runs on a local Mac computer for testing the applications (Yamacli, 2018).

Once an application is designed, it can be tested directly on the device or the simulator. The iOS Package App Store (IPA) is needed to distribute iOS applications. The IPA file includes a binary for the ARM architecture. The structure of the IPA file contains the payload folder, the iTunes Artwork file, iTunesMetadata.plist, WatchKitSupport/WK, and the application.app file. IPA files cannot be directly installed on iOS devices. Apple employs Fairplay Digital Rights Management (DRM), which works with public and private keys to manage digital rights and prevent the installation of applications on unauthorized devices.

An Apple ID is needed to deploy the application to an iPhone or iPad or to download software from the Apple store. Once a developer is logged in, Apple generates a pair of public and private keys for the username. The private key is delivered to the developer, and Apple keeps the public key. Then, a unique identifier is sent to the Apple server to register the mobile device. Therefore, if an attempt to run an IPA on a different device, a crash will succeed because the private key cannot decrypt the header encrypted in the public key.

A membership to the Apple Developer Program is needed to distribute applications on the App Store. Membership includes access to beta OS releases, advanced app capabilities, and tools to develop, test, and distribute apps. Once an application is uploaded to the store, Apple conducts an exhaustive review process before it can be published.

5.6   USER EXPERIENCE

An end-user can download and install a native application from the app store or directly onto the mobile device. In addition to correct operation, the end-user will prefer those applications that are accessible, convenient, and easy to use. User experience (UX) refers to the emotions and feelings that a customer experiences when using a software application. If users have a good experience with the application, they will use it continuously. Otherwise, they will feel discouraged and avoid using it again (Hassenzahl & Tractinsky, 2006).

According to Law et al. (2014), user experiences can be studied from two points of view: (i) the qualitative and (ii) the quantitative. Quantitative research yields numerical data to understand the statistical significance of the experience, and qualitative research yields non-numerical data in the form of opinions or feelings to understand the attitudes and behaviors of the users. However, measuring UX is complex because the application features, user context, and internal emotions influence it.

User experience in AR is spatial and highly contextual. Therefore, attention should be paid to the interaction and visual resources. In mobile AR, the factors that affect user experiences include the environment, the movement, the interaction, and the user interface. Chapter 6 explains the instruments to measure many aspects of an AR app.

Now that the steps to design an AR application have been explained, the reader will learn how to create the first AR experience using Meta Spark Studio in the following section.

5.7   CREATING AN AR PROJECT WITH META SPARK STUDIO

Meta Spark Studio allows the creation of effects and filters that can be shared on social networks. The reader will learn how to develop the first AR application using Meta Spark Studio in this section. One of the software's main features is the easy build, which is the reason why it was selected to design the first AR application.

The application will include a marker, which will be detected by the mobile device's camera using computer vision techniques that will be transparent to the designer. The marker will have a virtual model associated with it. The virtual model will be inserted into the real scene because of the detection and recognition of the marker. The marker can be moved, oriented, and scaled within the limits of the scene, thanks to the tracking techniques.

5.7.1   META SPARK STUDIO INSTALLATION

Before installing Meta Spark Studio, the reader must ensure that iTunes is installed. iTunes can be downloaded at https://support.apple.com/downloads/itunes. After downloading and installing iTunes, proceed to install Meta Spark Studio.

Meta Spark Studio can be downloaded at: https://sparkar.facebook.com/ar-studio/learn/downloads/#spark-ar-studio. The designer must select the corresponding operating system, Windows or macOS. Once the *.msi file is downloaded, proceed with the installation by clicking the file. A screen with the setup wizard is displayed. The end-user license agreement must be accepted. Next, the folder where the program will be installed must be defined. It is recommended to use the path offered by default. Then, the installation starts. An authentication process must be carried out with a Facebook account to use Meta Spark Studio.

In addition, the Meta Spark player for Android or iOS must be installed on the mobile device. This step must be conducted using the corresponding app store. At the time of writing this book, the Meta Spark Studio version available was v162. Regarding Meta Spak Player, the available version was 164.0. The iOS version was employed for this example.

5.7.2   THE FIRST AR APPLICATION

The following steps are needed to build the first AR app with Meta Spark Studio.

1. Open Meta Spark Studio by clicking the icon located on the computer's desktop.
2. The new screen shown in Figure 5.2 is displayed. On the left side, click on the menu "*Create New*" and then select "*Blank Project.*"

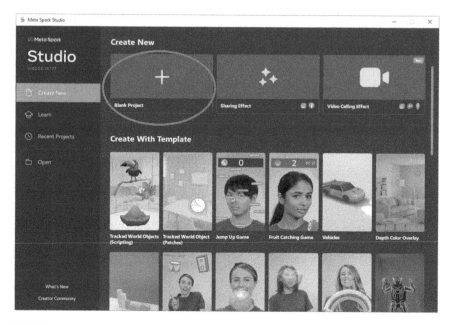

FIGURE 5.2   The screen to create the new project.

3. A new screen to design the project is displayed as shown in Figure 5.3. Please pay attention to the viewport showing the hotkeys (*A* allows to move left, *D* moves right). In the viewport, the effect that is being created can be designed and observed. Click the camera icon to select the image acquisition device in the lower-left corner. All the available cameras will be displayed. Please select one camera. Otherwise, select the "Real-time Simulation" in the middle of the window. One of the persons shown must be selected.

4. The video is displayed once the camera or simulation is selected. By default, the video is displayed vertically. However, the designer can change to the horizontal view by clicking the icon with an arrow, as shown in Figure 5.4. The vertical or horizontal orientation of the camera will be selected according to the designer's preferences.

5. In your favorite web browser, enter the following link: www.brosvision.com/ar-marker-generator/ to create the marker that will be used to activate the AR experience.

6. Select the shapes (lines, triangles, quadrangles) and quantities that will be inserted into the marker. Then push the button "Generate". For this example, a color marker will be generated. However, a grayscale marker can also serve. Once the marker is generated, please save it as "Marker.png" to your hard drive disk using the right-click. If you need more information regarding best practices for target tracking, please visit the link: https://sparkar.faceb ook.com/ar-studio/learn/articles/world-effects/best-practice-for-target-track ing#Initialize-tracking.

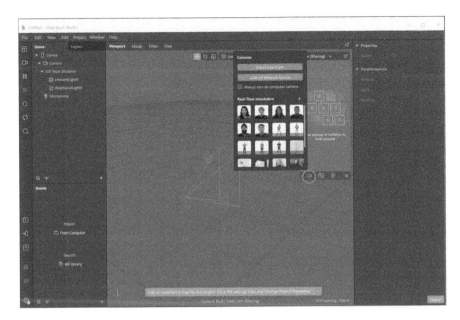

FIGURE 5.3 Screen for selecting the camera or the simulated environment.

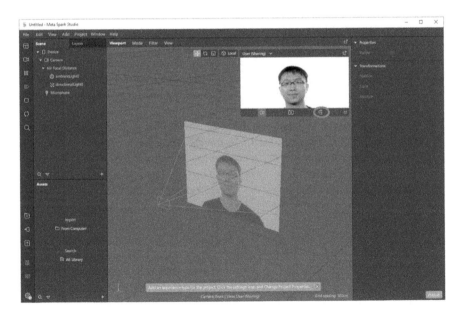

FIGURE 5.4 Changing the camera to the horizontal or vertical view.

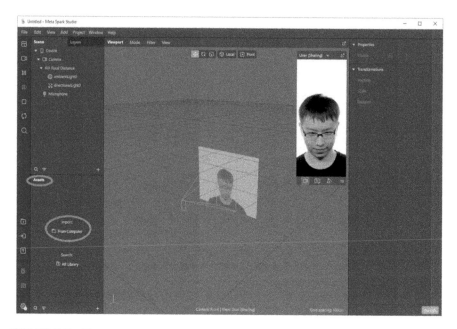

FIGURE 5.5 The steps to add a marker to the project.

7. Return to Meta Spark Studio.
8. The marker created in Brosvision must be inserted into the project. There are at least two techniques, as shown in Figure 5.5. (i) On the left bottom corner, detect the "Assets" space and then the "Import from Computer," push the button, navigate to your hard disk, and select the marker to be inserted, or (ii) drag the marker to the "Assets" area.
9. The marker will be inserted into the project under the menu Assets->Textures; this can take many seconds. Please be patient while the marker is compressing. In the end, the marker will be displayed as shown in Figure 5.6.
10. By clicking the "Plus" icon in the middle left or by entering the menu Add->Scene Understanding, insert a "Target Tracker" to the project. The "Target Tracker" will be inserted in the "Scene Space" and named "targetTracker0" as shown in Figure 5.7.
11. The project can contain many "Target Trackers." For this example, the project will contain only one "Target Tracker," as shown in Figure 5.8.
12. Select the "Target Tracker" in the scene tree, then select the menu "Texture" on the right panel to associate the marker inserted in step 9 to the "Texture" property. The marker will appear on the scene, as shown in Figure 5.9.
13. Following the instructions in step 8, insert the virtual model that will be displayed when the marker is recognized. The virtual model can be any 2D or 3D model. Alternatively, the "Search AR Library" menu can be used to select from many free resources, as shown in Figure 5.10.

FIGURE 5.6 The marker added to the project.

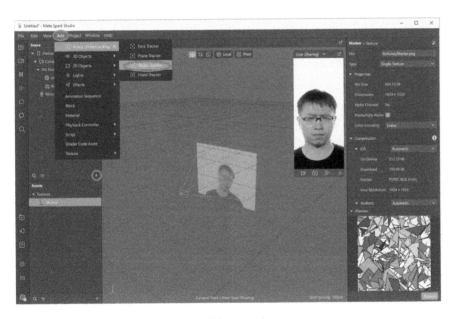

FIGURE 5.7 Adding the "Target Tracker" in the project.

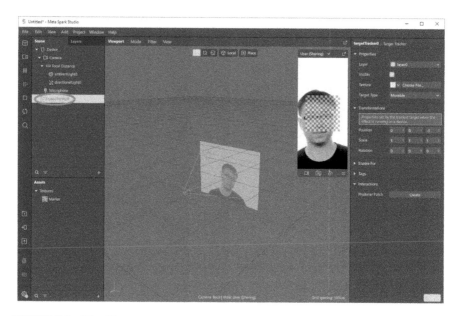

FIGURE 5.8   The "Target Tracker" added to the "Scene" space.

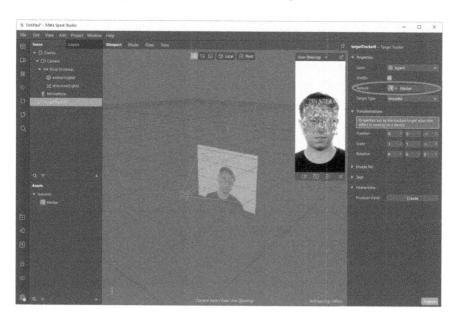

FIGURE 5.9   Adding the marker to the "Texture" property of the "Target Tracker."

14. The user will be asked to create a free "Sketchfab" account on the first usage, and then the free resources can be visited as shown in Figure 5.11.
15. The virtual model is inserted under the menu "Assets." In this example, a virtual model of a robot created by the book authors was inserted, as shown in Figure 5.12.

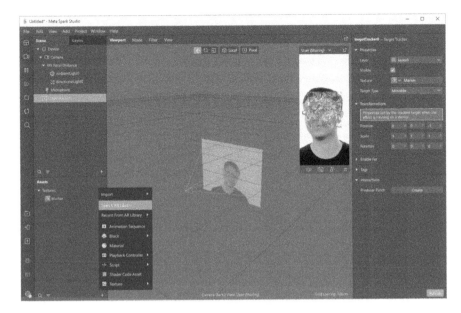

FIGURE 5.10 Inserting the virtual model.

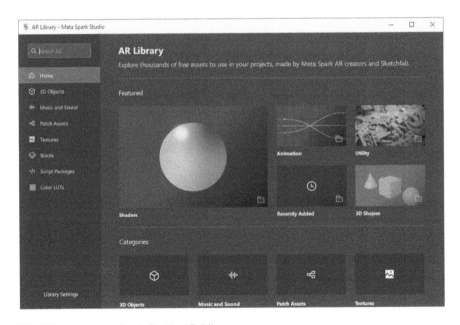

FIGURE 5.11 Meta Spark Studio AR Library.

16. Drag the virtual model to be part of the "Target Tracker." After the model appears on the scene, use the corresponding buttons to change the position, scale, and rotation properties to make the model look the way you want, as shown in Figure 5.13.

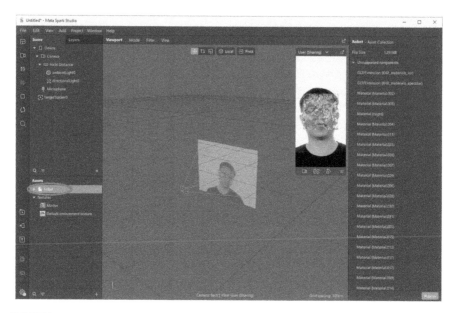

FIGURE 5.12 The virtual model inserted into the "Assets."

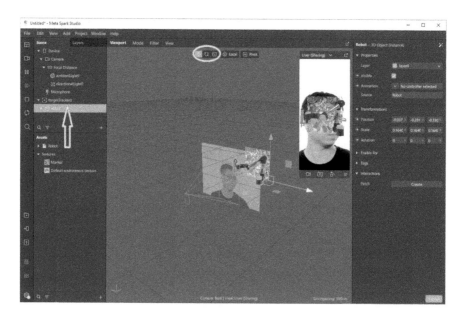

FIGURE 5.13 3D model inserted in the scene.

17. Save the project by following the path File->Save or clicking the shortcut control+s. It is recommended to save the project in the Meta Spark Studio folder. The project is saved as *.arproj. Moreover, a folder named equal to the project is generated. For this example, the project was saved as "ARBookExample."
18. The project has finished.

5.7.3 TESTING THE APPLICATION

The following steps must be executed for testing the application.

1. Connect the mobile device to a USB or a USB-C port. Confirm when asked about trusting the computer. For this example, an iOS mobile device was employed.
2. On the left menu, select "Test on Device" and click the button "Add Experience," as shown in Figure 5.14.
3. In the new screen, select the "Experiences" menu and click on the "Add Experience" button, as shown in Figure 5.15.
4. Select "Sharing Effect" on the new screen and click the "Insert" button, as shown in Figure 5.16.
5. A summary of the sharing effect is displayed. Then press the "Done" button, as shown in Figure 5.17.
6. Press the "Test on Device" button, and you will see a menu of the locations where the effect can be sent, including Instagram, Facebook, and mobile devices, as observed in Figure 5.18.
7. Open The "Meta Spak Player" on your mobile device.
8. Send the project to the mobile device, as shown in Figure 5.19.
9. The video of the back camera will be displayed.
10. Point the marker with the camera, and the virtual model will be superimposed in the scene.

The marker can be zoomed in or out and rotated to observe different views of the virtual model.

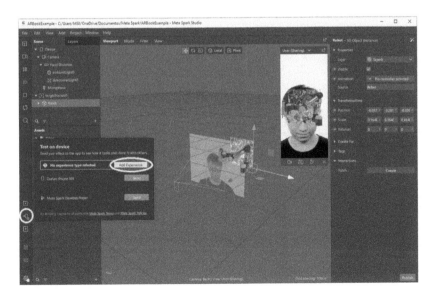

FIGURE 5.14   Creating the experience on the mobile device.

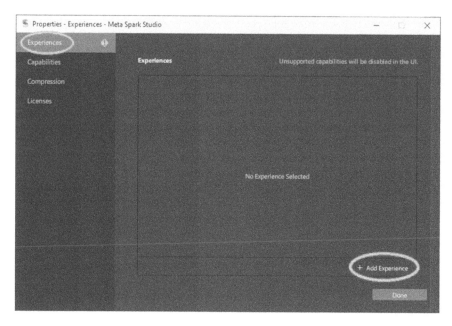

FIGURE 5.15   Adding the experience to the mobile device.

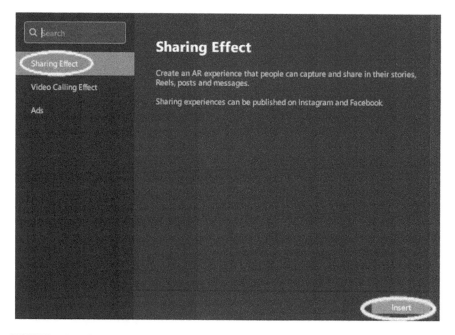

FIGURE 5.16   Sharing the AR effect.

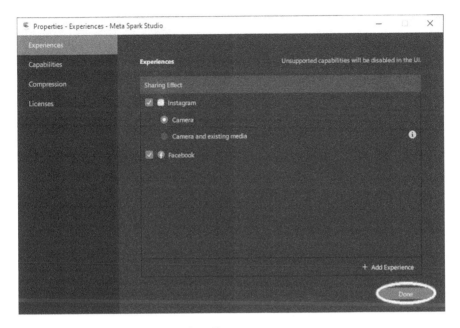

FIGURE 5.17   Summary of the sharing effect.

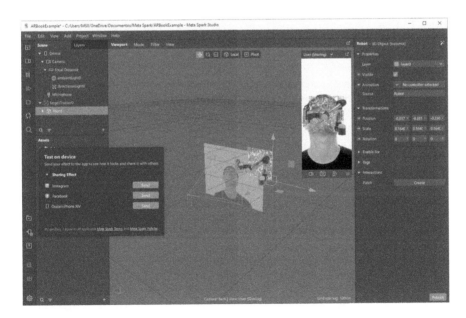

FIGURE 5.18   The summary of the applications where the AR can be sent.

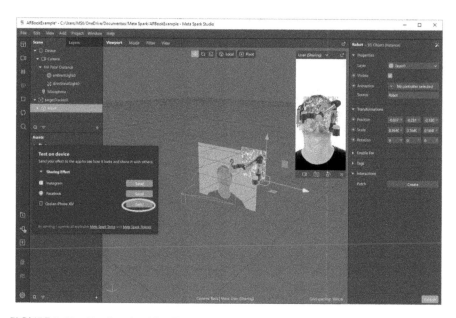

FIGURE 5.19   Sending the AR effect to the mobile device.

REFERENCES

Ableson, F., King, C., & Ortiz, C. (2011). *Android in Action*. Simon and Schuster.

Adobe. (2023, February). Adobe Aero. https://www.adobe.com/products/aero.html.

Afshar, J. (2023). *Hands-On Augmented Reality Development with Meta Spark Studio* (1st ed.). Apress.

Aumentaty. (2023, February). *Aumentaty Solutions*. http://www.aumentaty.com/index.php.

Blain, J. (2019). *The Complete Guide to Blender Graphics: Computer Modeling & Animation*. AK Peters/CRC Press.

Bollapragada, V., Murphy, C., & White, R. (2000). *Inside Cisco iOS Software Architecture*. Cisco Press.

Brosvision s.r.o. (2023, February). Augmented Reality Marker Generator – Brosvision. www. brosvision.com/ar-marker-generator/

Chopra, A. (2007). *Google SketchUp for Dummies*. John Wiley & Sons.

Chopra, A. (2012). *Introduction to Google Sketchup*. John Wiley & Sons.

Dearle, A. (2007). Software Deployment, Past, Present and Future. Proceeding of the Future of Software Engineering (FOSE) Conference, 269–284. https://doi.org/10.1109/ FOSE.2007.20

Deitel, P., & Deitel, H. (2016). *Android How to Program*. Pearson.

Feiler, J. (2014). *IOS App Development for Dummies*. John Wiley & Sons.

Fiala, M. (2010). Designing Highly Reliable Fiducial Markers. *IEEE Transactions on Pattern Analysis and Machine Intelligence*, *32*(7), 1317–1324. https://doi.org/10.1109/ TPAMI.2009.146

Fiorentino, M., Uva, A., Monno, G., & Radkowski, R. (2012). Augmented Technical Drawings: A Novel Technique for Natural Interactive Visualization of Computer-Aided Design Models. *Journal of Computing and Information Science in Engineering*, *12*(2), 24503. https://doi.org/10.1115/1.4006431

Hameed, Q., Hussein, H., Ahmed, M., & Basim, M. (2022). Development of Augmented Reality-Based Object Recognition Mobile Application with Vuforia. *Journal of Algebraic Statistics*, *13*(2), 2039–2046.

Hassenzahl, M., & Tractinsky, N. (2006). User Experience – A Research Agenda. *Behaviour & Information Technology*, *25*(2), 91–97. https://doi.org/10.1080/01449290500330331

Imbert, N., Vignat, F., Kaewrat, C., & Boonbrahm, P. (2013). Adding Physical Properties to 3D Models in Augmented Reality for Realistic Interactions Experiments. *Procedia Computer Science*, *25*, 364–369. https://doi.org/https://doi.org/10.1016/j.procs.2013.11.044

Javornik, A., Duffy, K., Rokka, J., Scholz, J., Nobbs, K., Motala, A., & Goldenberg, A. (2021). Strategic Approaches to Augmented Reality Deployment by Luxury Brands. *Journal of Business Research*, *136*, 284–292. https://doi.org/https://doi.org/10.1016/j.jbusres.2021.07.040

Kato, H., & Billinghurst, M. (1999). Marker Tracking and HMD Calibration for a Video-Based Augmented Reality Conferencing System. *Proceedings of the 2nd IEEE and ACM International Workshop on Augmented Reality (IWAR)*, 85–94. https://doi.org/10.1109/IWAR.1999.803809

Khan, R., Khan, S., Khan, H., & Ilyas, M. (2022). Systematic Literature Review on Security Risks and its Practices in Secure Software Development. *IEEE Access*, *10*, 5456–5481. https://doi.org/10.1109/ACCESS.2022.3140181

King, R. (2019). *3D Animation for the Raw Beginner Using Autodesk Maya 2e*. Chapman and Hall/CRC.

Krajci, I., & Cummings, D. (2013). History and Evolution of the Android OS. In Android on x86: An Introduction to Optimizing for Intel® Architecture (pp. 1–8). Apress. https://doi.org/10.1007/978-1-4302-6131-5_1

Lanham, M. (2018). *Learn ARCore-Fundamentals of Google ARCore: Learn to Build Augmented Reality Apps for Android, Unity, and the Web with Google ARCore 1.0*. Packt Publishing Ltd.

Law, E., van Schaik, P., & Roto, V. (2014). Attitudes Towards User Experience (UX) Measurement. *International Journal of Human-Computer Studies*, *72*(6), 526–541. https://doi.org/https://doi.org/10.1016/j.ijhcs.2013.09.006

Linowes, J., & Babilinski, K. (2017). *Augmented Reality for Developers: Build Practical Augmented Reality Applications with Unity, ARCore, ARKit, and Vuforia*. Packt Publishing Ltd.

Madden, L. (2011). *Professional Augmented Reality Browsers for Smartphones: Programming for Junaio, Layar and Wikitude*. John Wiley & Sons.

Marín, H., Alor, G., Colombo, L., Bustos, M., & Zatarraín, R. (2022). ZeusAR: A Process and an Architecture to Automate the Development of Augmented Reality Serious Games. *Multimedia Tools and Applications*, *81*(2), 2901–2935. https://doi.org/10.1007/s11042-021-11695-1

McQuilkin, K., & Powers, A. (2012). *Cinema 4D: The Artist's Project Sourcebook*. Routledge.

Mladenov, B., Damiani, L., Giribone, P., & Revetria, R. (2018). A Short Review of the SDKs and Wearable Devices to be Used for AR Application for Industrial Working Environment. *Proceedings of the World Congress on Engineering and Computer Science*, *1*, 23–25.

Murdock, K. (2022). *Autodesk Maya 2023 Basics Guide*. SDC Publications.

Permozer, I., & Orehovački, T. (2019). Utilizing Apple's ARKit 2.0 for Augmented Reality Application Development. Proceedings of the 42nd International Convention on Information and Communication Technology, Electronics and Microelectronics (MIPRO), 1629–1634. https://doi.org/10.23919/MIPRO.2019.8756928

Pooja, J., Vinay, M., Vineetha, G., & Anuradha, M. (2020). Comparative Analysis of Marker and Marker-less Augmented Reality in Education. Proceedings of the IEEE International

Conference for Innovation in Technology (INOCON), 1–4. https://doi.org/10.1109/INOCON50539.2020.9298303

Simonetti, A., & Paredes, J. (2013). *Vuforia v1. 5 SDK: Analysis and Evaluation of Capabilities*. Universidad Politécnica de Cataluña.

Szabo, M. (2012). *Cinema 4D R13 Cookbook*. Packt Publishing.

Tracy, K. (2012). Mobile Application Development Experiences on Apple's iOS and Android OS. *IEEE Potentials, 31*(4), 30–34. https://doi.org/10.1109/MPOT.2011.2182571

Usman, M., Felderer, M., Unterkalmsteiner, M., Klotins, E., Mendez, D., & Alégroth, E. (2020). Compliance Requirements in Large-Scale Software Development: An Industrial Case Study. In Torchiano, M., & Morisio Maurizio, J. A. (Eds.), *Product-Focused Software Process Improvement* (pp. 385–401). Springer International Publishing.

Van Gumster, J. (2020). *Blender for dummies*. John Wiley & Sons.

Wang, W. (2018). *Beginning ARKit for iPhone and iPad: Augmented Reality APP Development for iOS*. Apress.

Ximo, C. (2012). *ArToolKit*. Ject Press. https://books.google.com.mx/books?id=7fl8tgAACAAJ

Yamacli, S. (2018). *Beginner's Guide to iOS 12 App Development Using Swift 4: Xcode, Swift and App Design Fundamentals*. CreateSpace Independent Publishing Platform.

6 Instruments to Evaluate Augmented Reality

6.1 INTRODUCTION

Many scientists have asserted that AR is a helpful technology for education, medicine, robotics, entertainment, manufacturing, and training (Dargan et al., 2023; Villagran et al., 2023). However, it is not enough to only say that AR is useful; it is desirable that this can be demonstrated both qualitatively and quantitatively.

Once an AR prototype has been created, gaining insight into how users will accept it and the effect it will cause is recommendable. Evaluations are necessary to assess how well AR applications address users' needs. Moreover, user feedback from evaluations is essential to optimize revenues from these applications. The selection of an evaluation instrument depends on what needs to be evaluated and the hardware and software used.

Frequently, a questionnaire is employed to obtain user feedback regarding an AR product. A questionnaire is a self-reporting data collection instrument where the items are usually closed-ended and presented as multiple-choice. The respondents completed the questionnaire by selecting a set of alternatives or points on a rating scale (Marshall, 2005).

6.2 INSTRUMENTS EMPLOYED IN AR

Various instruments have been employed in the literature to measure different aspects of an AR prototype, including user achievement, effectiveness, technology acceptance, motivation, immersion, quality, usability, ergonomics, cognitive workload, satisfaction, task completion time, costs, and usefulness.

An overview of the approaches for evaluating AR can be consulted in the papers by Koutromanos and Kazakou (2023), Tzortzoglou and Sofos (2023), and Zigart and Schlund (2020). A brief description of various instruments employed in AR is presented in the following subsections. The original versions of the surveys are shown. However, the designer can modify each survey to adapt to their needs.

DOI: 10.1201/9781003435198-6

6.2.1 Pre-Test/Post-Test

The pre-test/post-test is one of the simplest quantitative evaluation models to compare groups and measure change resulting from experimental interventions or treatments. The pre-test is an evaluation applied before using the AR app to measure how much knowledge a person has regarding a particular theme. The post-test is an evaluation applied after using the AR app. With the information obtained from both tests, a comparison is conducted to measure the participants' knowledge change before and after using the AR app (Marsden & Torgerson, 2012).

A pre-test/post-test can be conducted using an experimental or quasi-experimental design. In experimental designs, the participants are randomly assigned to either receive the intervention (treatment group) or not (control group). Therefore, both groups participate in the pre-test. Then, the treatment group receives the treatment procedure, and the control group participates in a standard procedure. Finally, the post-test is administered to both groups. This study is frequently employed in controlled trials (van Riesen et al., 2022).

On the other hand, a quasi-experimental design is characterized because the sample to study is not selected randomly, and control groups are not required. Instead, participants are assigned to the sample based on non-random criteria previously established. Therefore, the pre-test is administered to the participants. Then, the treatment is executed, seeking to change the individual's score. Finally, the post-test is administered. This study is also called a nonrandomized or pre-post intervention and is frequently employed to conduct research in the educative field (Otte et al., 2019).

The pre-test/post-test design has been used in AR to measure achievements regarding laboratory skills (Akçayır et al., 2016), mechanics (Hedenqvist et al., 2023), vocabulary (Yilmaz et al., 2022), and financial mathematics (Hernández et al., 2021),

6.2.2 Technology Acceptance Model (TAM)

Davis, 1989 developed the Technology Acceptance Model (TAM) to explain how to encourage users to accept new technologies. TAM was derived from the psychology theory of reasoned action and the theory of planned behavior. Therefore, TAM suggested that the Perceived Ease of Use (PEU) and the Perceived Usefulness (PU) are determinants to explain what causes the behavioral Intention of a person To Use (ITU) a technology.

The Perceived Ease of Use (PEU) refers to the degree to which a person believes using a system would be free from effort. Perceived Usefulness (PU) refers to the degree to which the user believes a system would improve his/her work performance. Finally, the behavioral Intention To Use (ITU) measures the degree of technology acceptance (Marangunić & Granić, 2015).

TAM suggests that using a system is a response that can be explained or predicted by the user's motivation and is directly influenced by an external stimulus of the system's features and capabilities. The TAM survey comprises 14 seven-point Likert scale items. Therefore, the minimum score on the survey is 14, and the maximum is 98. The items of the TAM survey are shown in Table 6.1.

TABLE 6.1
Technology Acceptance Model (TAM) Survey

Please select the number that best represents how you feel about the AR prototype acceptance: 1 = Extremely disagree, 2 = Quite Disagree, 3 = Slightly Disagree, 4 = Neutral, 5 = Slightly Agree, 6 = Quite Agree, 7 = Extremely Agree.

	1	2	3	4	5	6	7
Perceived Usefulness (PU)							
PU1. Using the AR prototype in my job would enable me to accomplish tasks more quickly.							
PU2. Using the AR prototype would improve my job performance.							
PU3. Using the AR prototype in my job would increase my productivity.							
PU4. Using the AR prototype would enhance my effectiveness on the job.							
PU5. Using the AR prototype would make it easier to do my job.							
PU6. I would find the AR prototype useful in my job.							
Perceived Ease of Use (PEU)							
PEU1. Learning to operate the AR prototype would be easy for me.							
PEU2. I would find it easy to get the AR prototype to do what I want it to do.							
PEU3. My interaction with the AR prototype would be clear and understandable.							
PEU4. I would find the AR prototype to be flexible to interact with.							
PEU5. It would be easy for me to become skillful at using the AR prototype.							
PEU6. I would find the AR prototype easy to use.							
Intention to Use SICMAR (ITU)							
ITU1. I have the intention to use the AR prototype regularly for my job.							
ITU2. I have the intention to recommend the AR prototype.							

TAM was used in the literature to examine the adoption of AR technology in museum visiting experiences (Cheng et al., 2023), in a dance training system (Iqbal & Sidhu, 2022), in consumer perceptions (Oyman et al., 2022), and in learning the Mayo language (Miranda et al., 2016).

6.2.3 MOTIVATION

Motivation affects what, how, and when learners learn, and it is directly related to the development of students' attitudes and persistent efforts toward achieving a goal (Lin et al., 2021). Keller's Attention, Relevance, Confidence, and Satisfaction (ARCS) instructional model provides guidelines for designing and developing strategies to motivate students' learning (Keller, 2010a). Accordingly, motivation is an activity that must be performed to: (i) attract and sustain students' attention (A); (ii) define the relevance (R) of content students need to learn; (iii) help students believe they succeed in making efforts [gain confidence (C)]; and (iv) assist students in obtaining a sense of satisfaction (S) about their accomplishments in learning (Li & Keller, 2018).

The Instructional Materials Motivation Survey (IMMS) assesses students' motivation regarding a particular course based on the ARCS model and includes 36 items distributed as 12 items for (A), nine items for (R), nine items for (C), and six items for (S) (Keller, 2010b). The 36 items of the IMSS are shown in Table 6.2. The IMMS survey can be scored for each subscale or the total. The response scale ranges from 1 to 5, with a minimum score of 36 and a maximum of 180. However, each subscale can be used and scored independently.

The scores are computed by adding the responses for each subscale and the total scale. For the items marked as reverse, the responses should be changed to $5 = 1$, $4 = 2$, $3 = 3$, $2 = 4$, and $1 = 5$. The problem is that IMMS is long, and not all items are necessary. Therefore, the Reduced Instructional Materials Motivation Survey (RIMMS) is frequently employed to assess students' motivation.

RIMMS comprises 12 five-point Likert scale items, three for each ARCS dimension. The minimum score on the RIMMS survey is 12, and the maximum is 60, with a midpoint of 36 (Loorbach et al., 2014). The 12 items for RIMMS are shown in Table 6.3.

The IMMS has been used in AR studies to measure motivation in learning visual arts (Di Serio et al., 2013), math (Bhagat et al., 2021), and health education (Lin et al., 2021). In contrast, RIMMS has been used to measure motivation in learning Coulomb forces (Tomara & Gouscos, 2019), statistics (Gonzalez & Rosales, 2020), and aviation (Meister et al., 2023).

6.2.4 AUGMENTED REALITY IMMERSION (ARI)

Immersion is considered one of the main driving forces that promotes student learning in technological educational environments. According to Shin (2019), immersion is a perception of being physically present in a nonphysical world. Moreover, Dede (2009) stated that immersion helped to enhance science education in three ways: (i) multiple and complementary insights into complex scientific phenomena, (ii) situated learning, and (iii) the transfer of skills in real-world situations. In AR, immersion is the user's sensation of involvement in the AR world.

The Augmented Reality Immersion (ARI) questionnaire measures participants' perceived immersion using an AR prototype (Georgiou & Kyza, 2017). The ARI questionnaire is categorized into three scales: (i) engagement, (ii) engrossment, and (iii) total immersion. Engagement is divided into interest and usability. Engrossment

TABLE 6.2
Items of the Instructional Materials Motivation Survey (IMMS)

Please consider each statement concerning the instructional materials you have just studied and indicate how true it is. Give the answer that truly applies to you, not what you would like to be true or what you think others want to hear. A = Not True, B = Slightly True, C = Moderately True, D = Mostly True, and E = Very True.					
	A	**B**	**C**	**D**	**E**
C1. When I first looked at this lesson, I had the impression that it would be easy for me.					
A1. There was something interesting at the beginning of this lesson that got my attention.					
C2. This material was more difficult to understand than I would like for it to be. (reverse)					
C3. After reading the introductory information, I felt confident that I knew what I was supposed to learn from this lesson.					
S1. Completing the exercises in this lesson gave me a satisfying feeling of accomplishment.					
R1. It is clear to me how the content of this material is related to things I already know.					
C4. Many of the pages had so much information that it was hard to pick out and remember the important points. (reverse)					
A2. These materials are eye-catching.					
R2. There were stories, pictures, or examples that showed me how this material could be important to some people.					
R3. Completing this lesson successfully was important to me.					
A3. The quality of the writing helped to hold my attention.					
A4. This lesson is so abstract that it was hard to keep my attention on it. (reverse)					
C5. As I worked on this lesson, I was confident that I could learn the content.					
S2. I enjoyed this lesson so much that I would like to know more about this topic.					
A5. The pages of this lesson look dry and unappealing. (reverse)					

(continued)

TABLE 6.2 (Continued)
Items of the Instructional Materials Motivation Survey (IMMS)

Please consider each statement concerning the instructional materials you have just studied and indicate how true it is. Give the answer that truly applies to you, not what you would like to be true or what you think others want to hear. A = Not True, B = Slightly True, C = Moderately True, D = Mostly True, and E = Very True.

	A	B	C	D	E
R4. The content of this material is relevant to my interests.					
A6. The way the information is arranged on the pages helped keep my attention.					
R5. There are explanations or examples of how people use the knowledge in this lesson.					
C6. The exercises in this lesson were too difficult. (reverse)					
A7. This lesson has things that stimulated my curiosity.					
S3. I really enjoyed studying this lesson.					
A8. The amount of repetition in this lesson caused me to get bored sometimes. (reverse)					
R6. The content and style of writing in this lesson convey the impression that its content is worth knowing.					
A9. I learned some things that were surprising or unexpected.					
C7. After working on this lesson for a while, I was confident that I would be able to pass a test on it.					
R7. This lesson was not relevant to my needs because I already knew most of it. (reverse)					
S4. The wording of feedback after the exercises, or of other comments in this lesson, helped me feel rewarded for my effort.					
A10. The variety of reading passages, exercises, illustrations, etc., helped keep my attention on the lesson.					
A11. The style of writing is boring. (reverse)					
R8. I could relate the content of this lesson to things I have seen, done, or thought about in my own life.					

TABLE 6.2 (Continued)
Items of the Instructional Materials Motivation Survey (IMMS)

	A	B	C	D	E
Please consider each statement concerning the instructional materials you have just studied and indicate how true it is. Give the answer that truly applies to you, not what you would like to be true or what you think others want to hear. A = Not True, B = Slightly True, C = Moderately True, D = Mostly True, and E = Very True.					
A12. There are so many words on each page that it is irritating. (reverse)					
S5. It felt good to successfully complete this lesson.					
R9. The content of this lesson will be useful to me.					
C8. I could not really understand quite a bit of the material in this lesson. (reverse)					
C9. The good organization of the content helped me be confident that I would learn this material.					
S6. It was a pleasure to work on such a well-designed lesson.					

includes emotional attachment and focus of attention. Finally, total immersion comprises presence and flow. ARI comprises 21 seven-point Likert scale items. The minimum score on the survey is 21, and the maximum is 147. The items of the ARI questionnaire are shown in Table 6.4.

Readers should consult the works by Greenslade et al. (2022), Uriarte et al. (2022), and Zhang and Robb (2021) for examples of using the ARI survey in AR.

6.2.5 SOFTWARE QUALITY

Software quality describes the desirable characteristics of a software product. Attributes such as design, usability, operability, security, compatibility, maintainability, and functionality can be considered to define software quality (Dalla et al., 2020). However, questions such as how fast the system responds, how difficult it is to manipulate the system and markers, and to what extent the illumination affects marker recognition must be answered to measure the quality of an AR prototype.

The quality survey was designed by Barraza et al. (2015). The survey comprises 10 five-point Likert scale items. The minimum score on the survey is 10, and the maximum is 50. The items of the software quality survey are shown in Table 6.5.

The research by Hernández et al. (2021) can be consulted as an example of using the software quality survey in AR prototypes.

TABLE 6.3

Items of the Reduced Instructional Materials Motivation Survey (RIMMS)

Please think about each statement concerning the instructional materials you have just studied and indicate how true it is. Give the answer that truly applies to you, not what you would like to be true, or what you think others want to hear. A = Not True, B = Slightly True, C = Moderately True, D = Mostly True, and E = Very True.					
Attention	**A**	**B**	**C**	**D**	**E**
A1. The quality of the writing helped to hold my attention.					
A2. The way the information is arranged on the pages helped keep my attention.					
A3. The variety of reading passages, exercises, illustrations, etc., helped keep my attention on the lesson.					
Relevance					
R1. It is clear to me how the content of this material is related to things I already know.					
R2. The content and style of writing in this lesson convey the impression that its content is worth knowing.					
R3. The content of this lesson will be useful to me.					
Confidence					
C1. As I worked on this lesson, I was confident that I could learn the content.					
C2. After working on this lesson for a while, I was confident that I would be able to pass a test on it.					
C3. The good organization of the content helped me be confident that I would learn this material.					
Satisfaction					
S1. I enjoyed this lesson so much that I would like to know more about this topic.					
S2. I really enjoyed studying this lesson.					
S3. It was a pleasure to work on such a well-designed lesson.					

TABLE 6.4

Items of the Augmented Reality Immersion (ARI) Questionnaire

Please select the number that best represents how you feel regarding immersion using the AR prototype: 1 = Extremely disagree, 2 = Quite Disagree, 3 = Slightly Disagree, 4 = Neutral, 5 = Slightly Agree, 6 = Quite Agree, 7 = Extremely Agree.								
Immersion level	**Scale**	**1**	**2**	**3**	**4**	**5**	**6**	**7**
Engagement	**Interest**							
	I1. I liked the activity because it was novel.							
	I2. I liked the type of activity.							
	I3. I wanted to spend the time to complete the activity successfully.							
	I4. I wanted to spend time participating in the activity.							
	Usability							
	U1. It was easy for me to use the AR application.							
	U2. I found the AR application confusing.							
	U3. The AR application was unnecessarily complex.							
	U4. I did not have difficulties in controlling the AR application.							
Engrossment	**Emotional Attachment**							
	EA1. I was curious about how the activity would progress.							
	EA2. I was often excited since I felt part of the activity.							
	EA3. I often felt suspense by the activity.							
	Focus of attention							
	FA1. If interrupted, I looked forward to returning to the activity.							

(continued)

TABLE 6.4 (Continued)
Items of the Augmented Reality Immersion (ARI) Questionnaire

Please select the number that best represents how you feel regarding immersion using the AR prototype: 1 = Extremely disagree, 2 = Quite Disagree, 3 = Slightly Disagree, 4 = Neutral, 5 = Slightly Agree, 6 = Quite Agree, 7 = Extremely Agree.								
Immersion level	**Scale**	**1**	**2**	**3**	**4**	**5**	**6**	**7**
	FA2.Everyday thoughts and concerns faded out during the activity.							
	FA3. I was more focused on the activity rather than on any external distractions.							
Total Immersion	**Presence**							
	P1. The activity felt so authentic that it made me think that the virtual characters/objects existed for real.							
	P2. I felt that what I was experiencing was something real instead of a fictional activity.							
	P3. I was so involved in the activity that I sometimes wanted to interact directly with the virtual characters/objects.							
	P4. I was so involved that I felt that my actions could affect the activity.							
	Flow							
	F1. I did not have any irrelevant thoughts or external distractions during the activity.							
	F2. The activity became the unique and only thought occupying my mind.							
	F3. I lost track of time as if everything just stopped, and the only thing I could think about was the activity.							

TABLE 6.5
Items of the Software Quality Survey

Please select the number that best represents how you feel about the AR prototype quality: 1 = Not at all, 2 = A little, 3 = Moderately, 4 = Much, 5 = Very much.					
Quality questions	**1**	**2**	**3**	**4**	**5**
Q1. The AR prototype showed all the concepts explained by the teacher.					
Q2. The results obtained with the AR prototype were correct.					
Q3. The colors used for the AR prototype were adequate.					
Q4. The texts and numbers displayed by the AR prototype were legible.					
Q5. The size of the buttons allowed the easy manipulation of the AR prototype.					
Q6. The AR prototype velocity of response to carry out the calculations was fast.					
Q7. The classroom illumination was adequate.					
Q8. The manipulation of the electronic device I used was straightforward.					
Q9. Markers' manipulation was easy.					
Q10. The manipulation of the device in conjunction with the markers was easy.					

6.2.6 SYSTEM USABILITY SCALE (SUS)

The ease with which something can be used is known as usability. Moreover, usability can be described as the amount to which users can use a product (hardware-software) or service in a specific context of use (Keenan et al., 2022).

The System Usability Scale (SUS) is a metric employed to measure a system's usability perception rapidly. SUS was developed by Brooke (1996) and allowed for evaluating effectiveness, efficiency, and satisfaction. Effectiveness determines whether users were able to achieve their goals successfully. Efficiency calculates how much effort is needed to achieve the objectives. Finally, satisfaction measures the user's subjective reactions when using the system.

The SUS survey includes 10 questions that must be rated on a Likert scale from 1 to 5 (1: strongly disagree, 2: disagree, 3: neutral, 4: agree, and 5: strongly agree). The 10 items were selected from a pool of 50 items. The items cover aspects regarding system usability, such as training, complexity, and the need for support. Moreover,

TABLE 6.6
The System Usability Scale (SUS) Questionnaire

Please state your level of agreement or disagreement with the following statements based on your experience with the AR prototype. 1 = Strongly Disagree, 2 = Disagree, 3 = Neutral, 4 = Agree, and 5 = Strongly Agree.	1	2	3	4	5
U1. I think that I would like to use the AR prototype frequently.					
U2. I find the AR prototype unnecessarily complex.					
U3. I think the AR prototype was easy to use.					
U4. I think that I would need the support of a technical person to be able to use the AR prototype.					
U5. I find the various functions in the AR prototype were well integrated.					
U6. I think there was too much inconsistency in the AR prototype.					
U7. I imagine that most people would learn to use the AR prototype very quickly.					
U8. I find the AR prototype very cumbersome to use.					
U9. I feel very confident using the AR prototype.					
U10. I needed to learn a lot of things before I could get going with the AR prototype.					

the five odd questions (1, 3, 5, 7, and 9) are positive, and the five even questions (2, 4, 6, 8, and 10) are negative.

Therefore, in even-numbered questions, it is advisable to obtain the highest value; in odds, it is better to get the lowest score. The 10 questions of the SUS survey can be consulted in Table 6.6.

The algorithm to compute the SUS score is the following:

1. Sum up the total for the odd items (1, 3, 5, 7, 9).
2. Subtract five from the total to get the odd score.
3. Sum up the total for the even items (2, 4, 6, 8, 10).
4. Subtract the total from 25 to get an even score.
5. Sum up the even and odd scores.
6. Multiply the sum by 2.5 to obtain the final SUS score.

Although the SUS scores range from 0 to 100, they should not be considered a percentage. Therefore, a way to interpret the results is to normalize the scores to produce a percentile ranking. The SUS scale is divided into the following intervals: 1–51, 51–68, 68–80.3, +80.3, and results higher than 68 are considered good quality (Vlachogianni & Tselios, 2022). A result below 68 indicates that several aspects of the system must be corrected.

Currently, SUS is considered an industry standard, and its use can be found in the following articles Che et al., 2018; Daşdemir, 2022; Nazar et al., 2020; Thabit et al., 2022.

6.2.7 QUESTIONNAIRE FOR USER INTERACTION SATISFACTION (QUIS)

QUIS is a tool to evaluate the user's subjective rating regarding the AR interface. The questionnaire was presented in 1988 by a multidisciplinary team of researchers from the Human-Computer Interaction Laboratory of the University of Maryland (Chin et al., 1988).

The questionnaire effectively solved the problems of other instruments regarding standardization, reliability, and validity. QUIS includes five categories of questions, including: (i) overall reaction to the software, (ii) screen, (iii) terminology and system information, (iv) learning, and (v) system capabilities.

According to Chin et al. (1988) and Slaughter et al. (1995), the main goals of QUIS are: (i) to serve as a guide for systems design and redesign, (ii) to assess potential areas of system improvement, (iii) to provide a validated instrument to conduct comparative evaluations, and (iv) to serve as a test instrument in usability labs.

QUIS can evaluate systems and software in industrial and academic settings. Also, QUIS can quantify the magnitude of improvements before and after conducting changes to a system. QUIS 7.0 is the current version. QUIS version, which includes 27 nine-point Likert scale items, is shown in Table 6.7.

TABLE 6.7
Items of the Questionnaire for User Interaction Satisfaction (QUIS)

For each of the following questions, fill in 0–9 or leave blank if the question is not applicable.													
Overall Reaction to the Software			**0**	**1**	**2**	**3**	**4**	**5**	**6**	**7**	**8**	**9**	
QS1	Terrible												Wonderful
QS2	Difficult												Easy
QS3	Frustrating												Satisfying
QS4	Inadequate power												Adequate power
QS5	Dull												Stimulating

(continued)

TABLE 6.7 (Continued)
Items of the Questionnaire for User Interaction Satisfaction (QUIS)

For each of the following questions, fill in 0–9 or leave blank if the question is not applicable.													
Overall Reaction to the Software		**0**	**1**	**2**	**3**	**4**	**5**	**6**	**7**	**8**	**9**	
QS6	Rigid												Flexible
Screen													
S1. Reading characters on the screen.	Hard												Easy
S2. Highlighting simplifies task.	Not at all												Very much
S3. Organization of information.	Confusing												Very clear
S4. Sequence of screens.	Confusing												Very clear
Terminology and System Information													
T1. Use of terms throughout the system.	Inconsistent												Consistent
T2. Terminology related to task.	Never												Always
T3. Position of messages on screen.	Inconsistent												Consistent
T4. Prompts for input.	Confusing												Clear
T5. Computer informs about its progress.	Never												Always
T6. Error messages.	Unhelpful												Helpful

TABLE 6.7 (Continued)
Items of the Questionnaire for User Interaction Satisfaction (QUIS)

For each of the following questions, fill in 0–9 or leave blank if the question is not applicable.

Overall Reaction to the Software		0	1	2	3	4	5	6	7	8	9	
Learning													
L1. Learning to operate the system	Difficult											Easy
L2. Exploring new features by trial and error.	Difficult											Easy
L3. Remembering names and use of command	Difficult											Easy
L4. Performing tasks is straightforward	Never											Always
L5. Help messages on the screen	Unhelpful											Helpful
L6. Supplemental reference materials	Confusing											Clear
System Capabilities													
SC1. System speed	Too slow											Fast enough
SC2. System reliability	Unreliable											Reliable
SC3. System tends to be	Noisy											Quiet
SC4. Correcting your mistakes	Difficult											Easy
SC5. Designed for all levels of users	Never											Always

The reader can consult the works by Helin et al. (2018) and Xue et al. (2019) to gain insight into how QUIS has been employed in AR.

6.2.8 SMART GLASSES USER SATISFACTION (SGUS)

In some AR applications, it is imperative to keep the users' hands free to manipulate any artifact. Therefore, smart glasses are used instead of mobile devices to experiment with AR. However, the experience of using smart glasses is not always satisfactory. Consequently, the Smart Glasses User Satisfaction (SGUS) questionnaire was designed to assess users' subjective satisfaction with smart glasses (Helin et al., 2018).

SGUS was created for the Wearable Experience for Knowledge Intensive Training (WEKIT) trials. WEKIT is a European project to develop and test a novel way of industrial training enabled by smart wearable technology (Klemke et al., 2017). The general objective of the questionnaire is to understand the potential end users' central expectations of AR services with smart glasses.

SGUS was created by mixing and modifying the evaluation criteria proposed by Ssemugabi and De Villiers (2007) and statements taken from Olsson (2013). SGUS consists of 11 items (statements) with a seven-point Likert scale. The summation of the score for the 11 items is the SGUS score. The 11 statements include three evaluation criteria categories: general interface usability criteria, AR interaction-specific criteria for an educational AR app, and learner-centered effective learning (Ssemugabi & De Villiers, 2007). The items of the SGUS questionnaire are shown in Table 6.8.

The reader can consult the papers by Helin et al. (2018) and Xue et al. (2019) to see examples of using SGUS in AR applications.

6.2.9 USEFULNESS, SATISFACTION, AND EASE OF USE (USE)

The Usefulness, Satisfaction, and Ease of Use (USE) questionnaire was developed in 2001 by Lund and colleagues from Ameritech, U.S. WEST Advanced Technologies, and Sapient to measure a product's or service's subjective usability. USE consists of 30 items on a seven-point Likert scale that examines four dimensions of usability: (i) usefulness, (ii) ease of use, (iii) ease of learning, and (iv) satisfaction (Lund, 2001).

USE was created from a large pool of items collected from internal studies, literature, and brainstorming. The main goal of USE was to make the items as simply worded and general as possible. Therefore, USE is a complete and very simple to implement questionnaire. Users are asked to rate agreement with the statements, ranging from strongly disagree (lowest score) to strongly agree (maximum score).

USE can be applied to various usability scenarios because it is non-proprietary and technology-agnostic. USE also has the possibility of adapting the questionnaire questions to particular needs. USE items have good face validity with unambiguous and relevant descriptions (Gao et al., 2018). The items of the USE questionnaire are shown in Table 6.9.

Albertazzi et al. (2012) employed the USE questionnaire to measure the usability when a person first interacts with a product using AR. Moreover, the reader can

TABLE 6.8
Items of the Smart Glasses User Satisfaction (SGUS) Questionnaire

Please select the number that best represents how satisfied you feel when using smart glasses for testing the AR prototype: 1 = Extremely disagree, 2 = Quite Disagree, 3 = Slightly Disagree, 4 = Neutral, 5 = Slightly Agree, 6 = Quite Agree, 7 = Extremely Agree.							
	1	**2**	**3**	**4**	**5**	**6**	**7**
SG1. With AR-glasses I could access information at the most appropriate place and moment.							
SG2. Content displayed on the AR-glasses made sense in the context I used it.							
SG3. AR-glasses provided me with the most suitable amount of information.							
SG4. AR-glasses allowed a natural way to interact with information displayed.							
SG5. I had a good conception of what is real and what is augmented when using AR-glasses.							
SG6. The interaction with content on AR-glasses captivated my attention in a positive way.							
SG7. The instructions given by AR-glasses helped me to accomplish the task.							
SG8. I understood what is expected from me in each phase of the task with the help of AR-glasses.							
SG9. Performing the task with the help of AR-glasses was natural to me.							
SG10. While using AR-glasses, I was aware of the phase of the task at all times during the execution of the task.							
SG11. While using AR-glasses, I was able to pay attention to the essential aspects of the task all the time.							

consult other examples that employed USE in AR settings for tourism (Ramli et al., 2021) and learn idioms (Edyanto et al., 2021) and history (Ibharim et al., 2021).

6.2.10 SOFTWARE USABILITY MEASUREMENT INVENTORY (SUMI)

The Software Usability Measurement Inventory (SUMI) is a payment service rigorously tested and proven for measuring software or service usability from the end

TABLE 6.9
Items of Usefulness, Satisfaction, and Ease of Use (USE) Survey

Please rate your agreement with these statements. Try to respond to all items. For items that are not applicable, use NA. 1 = Strongly Disagree, 2 = Quite Disagree, 3 = Slightly Disagree, 4 = Neutral, 5 = Slightly Agree, 6 = Quite Agree, 7 = Strongly Agree.								
Usefulness	1	2	3	4	5	6	7	N/A
U1. It helps me be more effective.								
U2. It helps me be more productive.								
U3. It is useful.								
U4. It gives me more control over the activities in my life.								
U5. It makes the things I want to accomplish easier to get done.								
U6. It saves me time when I use it.								
U7. It meets my needs.								
U8. It does everything I would expect it to do.								
Ease of Use								
EOU1. It is easy to use.								
EOU2. It is simple to use.								
EOU3. It is user friendly.								
EOU4. It requires the fewest steps possible to accomplish what I want to do with it.								
EOU5. It is flexible.								
EOU6. Using it is effortless.								
EOU7. I can use it without written instructions.								
EOU8. I don't notice any inconsistencies as I use it.								
EOU9. Both occasional and regular users would like it.								
EOU10. I can recover from mistakes quickly and easily.								
EOU11. I can use it successfully every time.								
Ease of Learning								
EOL1. I learned to use it quickly.								

TABLE 6.9 (Continued)
Items of Usefulness, Satisfaction, and Ease of Use (USE) Survey

Please rate your agreement with these statements. Try to respond to all items. For items that are not applicable, use NA. 1 = Strongly Disagree, 2 = Quite Disagree, 3 = Slightly Disagree, 4 = Neutral, 5 = Slightly Agree, 6 = Quite Agree, 7 = Strongly Agree.								
Usefulness	1	2	3	4	5	6	7	N/A
EOL2. I easily remember how to use it.								
EOL3. It is easy to learn to use it.								
EOL4. I quickly became skillful with it.								
Satisfaction								
S1. I am satisfied with it.								
S2. I would recommend it to a friend.								
S3. It is fun to use.								
S4. It works the way I want it to work.								
S5. It is wonderful.								
S6. I feel I need to have it.								
S7. It is pleasant to use.								

user's point of view. The work on SUMI started in 1990 as a part of the Metrics for Usability Standards in Computing (MUSiC) project from the Human Factors Research Group (HFRG), University College, Cork, Ireland. MUSiC is a project to develop methods for assessing usability. Therefore, SUMI can be used to assess new products, compare previous versions of products, and set targets for future application developments (Kirakowski & Corbett, 1993). SUMI has also been used to set user experience requirements by software procurers.

A minimum of 10 users is recommended to use SUMI effectively. An extensive reference database and an analysis and report generation tool called SUMI Scoring program (SUMISCO) support SUMI (Dumas & Redish, 1993).

SUMI is a method recognized by ISO 9241 standard for testing user satisfaction and comprises 50 items and five subscales for efficiency, affect, helpfulness, control, and learnability (10 items each) to be assessed on a scale from 1 to 3 (agree, undecided, and disagree). The mean score of SUMI is 50, with a standard deviation of 10. If a range between 40 and 60 is obtained, it can be affirmed that the product is usable. The items of the SUMI questionnaire are shown in Table 6.10.

Examples of the use of SUMI to measure the usability of AR prototypes can be found in the works of Campbell et al. (2014), Maier et al. (2009), and Pratama and Sukirman (2023).

TABLE 6.10
Items of the Software Usability Measurement Inventory (SUMI)

Please answer the 50 items. After each statement, there are three boxes. Check the first box if you generally AGREE with the statement, the middle if you are UNDECIDED or if the statement has no relevance to your software or to your situation, and the right if you generally DISAGREE with the statement. In checking the left or right box, you are not necessarily indicating strong agreement or disagreement but just your general feeling most of the time.			

	Agree	Undecided	Disagree
SUMI1. This software responds too slowly to inputs.			
SUMI2. I would recommend this software to my colleagues.			
SUMI3. The instructions and prompts are helpful.			
SUMI4. This software has at some time stopped unexpectedly.			
SUMI5. Learning to operate this software initially is full of problems.			
SUMI6. I sometimes don't know what to do next with this software.			
SUMI7. I enjoy the time I spend using this software.			
SUMI8. I find that the help information given by this software is not very useful.			
SUMI9. If this software stops, it is not easy to restart it.			
SUMI10. It takes too long to learn the software functions.			
SUMI11. I sometimes wonder if I am using the right function.			
SUMI12. Working with this software is satisfying.			
SUMI13. The way that system information is presented is clear and understandable.			
SUMI14. I feel safer if I use only a few familiar functions.			

TABLE 6.10 (Continued)
Items of the Software Usability Measurement Inventory (SUMI)

Please answer the 50 items. After each statement, there are three boxes. Check the first box if you generally AGREE with the statement, the middle if you are UNDECIDED or if the statement has no relevance to your software or to your situation, and the right if you generally DISAGREE with the statement. In checking the left or right box, you are not necessarily indicating strong agreement or disagreement but just your general feeling most of the time.

	Agree	Undecided	Disagree
SUMI15. The software documentation is very informative.			
SUMI16. This software seems to disrupt the way I normally like to arrange my work.			
SUMI17. Working with this software is mentally stimulating.			
SUMI18. There is never enough information on the screen when it's needed.			
SUMI19. I feel in command of this software when I am using it.			
SUMI20. I prefer to stick to the functions that I know best.			
SUMI21. I think this software is inconsistent.			
SUMI22. I would not like to use this software every day.			
SUMI23. I can understand and act on the information provided by this software.			
SUMI24. This software is awkward when I want to do something which is not standard.			
SUMI25. There is too much to read before you can use the software.			
SUMI26 Tasks can be performed in a straightforward manner using this software.			
SUMI27. Using this software is frustrating.			
SUMI28. The software has helped me overcome any problems I have had in using it.			
SUMI29. The speed of this software is fast enough.			

(continued)

TABLE 6.10 (Continued)
Items of the Software Usability Measurement Inventory (SUMI)

	Agree	Undecided	Disagree
Please answer the 50 items. After each statement, there are three boxes. Check the first box if you generally AGREE with the statement, the middle if you are UNDECIDED or if the statement has no relevance to your software or to your situation, and the right if you generally DISAGREE with the statement. In checking the left or right box, you are not necessarily indicating strong agreement or disagreement but just your general feeling most of the time.			
SUMI30. I keep having to go back to look at the guides.			
SUMI31. There have been times in using this software when I have felt quite tense.			
SUMI32. The organization of the menus seems quite logical.			
SUMI33. The software allows the user to be economic of keystrokes.			
SUMI34. Learning how to use new functions is difficult.			
SUMI35. There are too many steps required to get something to work.			
SUMI36. I think this software has sometimes given me a headache.			
SUMI37. Error messages are not adequate.			
SUMI38. It is easy to make the software do exactly what you want.			
SUMI39. I will never learn to use all that is offered in this software.			
SUMI40. There have been times in using this software when I have felt quite tense.			
SUMI41. The software hasn't always done what I was expecting.			
SUMI42. The software presents itself in a very attractive way.			
SUMI43. Either the amount or quality of the help information varies across the system.			
SUMI44 It is relatively easy to move from one part of a task to another.			

TABLE 6.10 (Continued)
Items of the Software Usability Measurement Inventory (SUMI)

Please answer the 50 items. After each statement, there are three boxes. Check the first box if you generally AGREE with the statement, the middle if you are UNDECIDED or if the statement has no relevance to your software or to your situation, and the right if you generally DISAGREE with the statement. In checking the left or right box, you are not necessarily indicating strong agreement or disagreement but just your general feeling most of the time.			
	Agree	**Undecided**	**Disagree**
SUMI45. It is easy to forget how to do things with this software.			
SUMI46. This software occasionally behaves in a way which can't be understood.			
SUMI47. This software is really very awkward.			
SUMI48. It is easy to see at a glance what the options are at each stage.			
SUMI49. Getting data files in and out of the system is not easy.			
SUMI50. I have to look for assistance most of the time when I use this software.			

6.2.11 Computer System Usability Questionnaire (CSUQ)

The Computer Systems Usability Questionnaire (CSUQ) is an instrument developed by IBM for measuring user satisfaction with system usability that is identical to the Post-Study System Usability Questionnaire (PSSUQ), except that the wording of the items does not refer to a usability testing situation. The PSSUQ uses the past tense, and the CSUQ uses the present tense. The CSUQ was designed to be a non-laboratory version of the PSSUQ (Lewis, 1995).

CSUQ is made up of 19 questions. Each question is a statement with a seven-point Likert scale from "Strongly Disagree" to "Strongly Agree." CSUQ measures four factors: (i) system usefulness, made up of items 1–6; (ii) information quality, made up of items 7–12; (iii) interface quality, made up of items 13–17; and (iv) overall satisfaction, made up of items 18 and 19 (Lewis, 2018).

CSUQ has presented high levels of reliability over time, which is evidence of stability in their internal consistency across the different versions. CSUQ can produce four scores, one overall score (average of 19 items), and three subscales. The 19 items of CSUQ are shown in Table 6.11.

The use of CSUQ in assessing the usability of AR prototypes can be consulted in the works by Caria et al. (2020), Criollo et al. (2021), and Korkut et al. (2022).

TABLE 6.11

Items of the Computer System Usability Questionnaire (CSUQ)

Items	1	2	3	4	5	6	7	N/A
The questionnaire allows you to express satisfaction with the usability of a system. Your responses will help us understand the aspects of the system you are particularly concerned about and the aspects that satisfy you. To as great a degree as possible, think about all the tasks you have done with the system while answering these questions. Please read each statement and indicate how strongly you agree or disagree with the statement by selecting a number on the scale. If a statement does not apply to you, select N/A. Thank you.								
System Usefulness								
CSUQ1. Overall, I am satisfied with how easy it is to use system.								
CSUQ2. It was simple to use system.								
CSUQ3. I can effectively complete my work using system.								
CSUQ4. I am able to complete my work quickly using system.								
CSUQ5. I am able to efficiently complete my work using system.								
CSUQ6. I feel comfortable using system.								
Information Quality								
CSUQ7. It was easy to learn to use system.								
CSUQ8. I believe I became productive quickly using system.								
CSUQ9. The system gives error messages that clearly tell me how to fix problems.								
CSUQ10. Whenever I make a mistake using system, I recover easily and quickly.								
CSUQ11. The information (such as online help, on-screen messages, and other documentation) provided with the system is clear.								
CSUQ12. It is easy to find the information I needed.								
Interface Quality								
CSUQ13. The information provided for the system is easy to understand.								

TABLE 6.11 (Continued)
Items of the Computer System Usability Questionnaire (CSUQ)

The questionnaire allows you to express satisfaction with the usability of a system. Your responses will help us understand the aspects of the system you are particularly concerned about and the aspects that satisfy you. To as great a degree as possible, think about all the tasks you have done with the system while answering these questions. Please read each statement and indicate how strongly you agree or disagree with the statement by selecting a number on the scale. If a statement does not apply to you, select N/A. Thank you.

Items	1	2	3	4	5	6	7	N/A
CSUQ14. The information is effective in helping me complete the tasks and scenarios.								
CSUQ15. The organization of information on system screens is clear.								
CSUQ16. The interface of the system is pleasant.								
CSUQ17. I like using the interface of this system.								
Overall Satisfaction								
CSUQ18. This system has all the functions and capabilities I expect it to have.								
CSUQ19. Overall, I am satisfied with the system.								

6.2.12 HANDHELD AUGMENTED REALITY USABILITY SCALE (HARUS)

The Handheld Augmented Reality Usability Scale (HARUS) evaluates perceptual and ergonomic issues encountered using handheld AR prototypes. HARUS was created due to a systematic literature review of the common problems reported by handheld AR users. The questionnaire comprises 16 items divided into two main sections regarding comprehensibility (ease of understanding the information presented by the handheld AR prototype) and manipulability (ease of handling the handheld AR prototype) (Santos et al., 2014).

Five stages were followed to create HARUS, including: (i) background study, (ii) questionnaire conceptualization, (iii) selection of the format and data analysis, (iv) establishing validity, and (v) establishing reliability. The survey is based on a seven-point Likert scale. The user must rate each statement from 1 to 7; only 1 (strongly disagree) and 7 (strongly agree) were labeled. Moreover, the questions were ordered in such a way as to alternate positively and negatively stated questions (Santos et al., 2015).

The HARUS questionnaire is shown in Table 6.12. The HARUS score is computed using a similar method to the SUS score.

1. Convert the scores to range from zero to six.
2. A one must be subtracted from the user response for the positive items.

TABLE 6.12
The Handheld Augmented Reality Usability Scale (HARUS) Questionnaire

For each of the following statements, mark one circle that best describes your reaction to the handheld augmented reality application. 1 = Strongly Disagree, 7 = Strongly Agree.							
Manipulability measures	**1**	**2**	**3**	**4**	**5**	**6**	**7**
M1. I think that interacting with this application requires a lot of body muscle effort.							
M2. I felt that using the application was comfortable for my arms and hands.							
M3. I found the device difficult to hold while operating the application.							
M4. I found it easy to input information through the application.							
M5. I felt that my arm or hand became tired after using the application.							
M6. I think the application is easy to control.							
M7. I felt that I was losing grip and dropping the device at some point.							
M8. I think the operation of this application is simple and uncomplicated.							
Comprehensibility measures							
C1. I think that interacting with this application requires a lot of mental effort.							
C2. I thought the amount of information displayed on the screen was appropriate.							
C3. I thought the information displayed on the screen was difficult to read.							
C4. I felt that the information display was responding fast enough.							
C5. I thought that the information displayed on the screen was confusing.							
C6. I thought the words and symbols on the screen were easy to read.							
C7. I felt that the display was flickering too much.							
C8. I thought that the information displayed on the screen was consistent.							

3. Subtract the user response from seven for the negative items.
4. Add all the converted responses.
5. Divide the sum by 0.96 to obtain a score of 0 to 100.

HARUS was designed to answer the following questions: (i) is this system easy to handle? and (ii) is the information presented easy to understand? Therefore, HARUS did not evaluate the handheld AR application overall. The internal consistency of HARUS was validated with an alpha Cronbach value of α = 0.83, which is a good value. Moreover, Harus must be used to measure how well users can use the handheld AR prototype.

The reader can review the works of Dutta et al. (2022), Fuvattanasilp et al. (2021), Polvi et al. (2016), and Santos et al. (2016) as examples of using the HARUS questionnaire to measure usability in mobile AR prototypes.

6.2.13 NASA TASK LOAD INDEX

The National Aeronautics and Space Administration (NASA) Task Load Index (NASA TLX) is one of the most used instruments to measure cognitive load using an AR prototype. The Human Performance Group developed NASA TLX as a result of a 3-year project. NASA TLX is a multidimensional assessment procedure that computes an overall workload score based on a weighted average of the scores on six dimensions: (i) mental demand (low/high), (ii) physical demand (low/high), (iii) temporal demand (low/high), (iv) performance (good/poor), (v) effort (low/high), and (vi) frustration level (low/high) (Hart & Staveland, 1988).

Mental demand explains how much mental and perceptual activity is required to conduct a specific task. Physical demand measures how much physical activity is required to conduct the task. Temporal demand describes how much time pressure the user feels due to the pace at which the tasks or task elements occur. Own performance explains how successful the user was in performing the task. Effort describes how hard the user has to work (mentally and physically) to accomplish a level of performance. Finally, the frustration level explains how irritated, stressed, and annoyed versus content, relaxed, and complacent the user feels during the task (Hart, 2006).

The NASA TLX instrument is applied in two phases: the weighting phase, which occurs just before the execution of the task, and the scoring phase, which occurs immediately after the execution. In the weighting phase, participants are presented with each dimension's definitions to choose an extreme value. A weight of 0 is assigned for the dimension that has not been chosen and 5 for the dimension that has been chosen the most.

In the scoring phase, participants rate the task they completed, marking a point for each dimension on the scale presented. Each dimension is presented on a line divided into 20 equal intervals. Then, the score is converted to a scale of 100. The items of the NASA TLX are shown in Table 6.13.

The works by Atici et al. (2021), Buchner et al. (2021), Condino et al. (2020), and Wang et al. (2023) can be consulted to gain insights about using NASA TLX in AR.

TABLE 6.13
Items of the NASA TLX

Task developed:
Mental demand: How mentally demanding was the task?
Very low   Very High
Physical demand: how physically demanding was the task?
Very low   Very High
Temporal demand: How hurried or rushed was the pace of the task?
Very low   Very High
Performance: How successful were you in accomplishing what you were asked to do?
Good   Poor
Effort: How hard did you have to work to accomplish your level of performance?
Very low   Very High
Frustration level: How insecure, discouraged, irritated, stressed, and annoyed were you?
Very low   Very High

REFERENCES

Akçayır, M., Akçayır, G., Pektaş, H., & Ocak, M. (2016). Augmented Reality in Science Laboratories: The Effects of Augmented Reality on University Students' Laboratory Skills and Attitudes Toward Science Laboratories. *Computers in Human Behavior*, 57(1), 334–342. https://doi.org/https://doi.org/10.1016/j.chb.2015.12.054

Albertazzi, D., Okimoto, M., & Ferreira, M. (2012). Developing an Usability Test to Evaluate the Use of Augmented Reality to Improve the First Interaction with a Product. *Work*, 41, 1160–1163. https://doi.org/10.3233/WOR-2012-0297-1160

Atici, H., Dila, Y., Taskapilioglu, O., & Tulin, G. (2021). Effects of Augmented Reality Glasses on the Cognitive Load of Assembly Operators in the Automotive Industry. *International*

Journal of Computer Integrated Manufacturing, *34*(5), 487–499. https://doi.org/ 10.1080/0951192X.2021.1901314

Barraza, R., Vergara, O., & Cruz, V. (2015). A Mobile Augmented Reality Framework Based on Reusable Components. *IEEE Latin America Transactions*, *13*(3), 713–720. https:// doi.org/10.1109/TLA.2015.7069096

Bhagat, K., Yang, F., Cheng, C., Zhang, Y., & Liou, W. (2021). Tracking the Process and Motivation of Math Learning with Augmented Reality. *Educational Technology Research and Development*, *69*(6), 3153–3178. https://doi.org/10.1007/s11423-021-10066-9

Brooke, J. (1996). Sus: A Quick and Dirty Usability. *Usability Evaluation in Industry*, *189*(3), 189–194.

Buchner, J., Buntins, K., & Kerres, M. (2021). A Systematic Map of Research Characteristics in Studies on Augmented Reality and Cognitive Load. *Computers and Education Open*, *2*, 1–8. https://doi.org/https://doi.org/10.1016/j.caeo.2021.100036

Campbell, A., Stafford, J., Holz, T., & O'Hare, G. (2014). Why, When and How to Use Augmented Reality Agents (AuRAs). *Virtual Reality*, *18*(2), 139–159. https://doi.org/ 10.1007/s10055-013-0239-4

Caria, M., Todde, G., Sara, G., Piras, M., & Pazzona, A. (2020). Performance and Usability of Smartglasses for Augmented Reality in Precision Livestock Farming Operations. *Applied Sciences*, *10*(7), 1–11. https://doi.org/10.3390/app10072318

Che, N., Abd, N., Arshad, H., & Khalid, W. (2018). User Satisfaction for an Augmented Reality Application to Support Productive Vocabulary Using Speech Recognition. *Advances in Multimedia*, *2018*(1), 1–10. https://doi.org/10.1155/2018/9753979

Cheng, A., Ma, D., Pan, Y., & Hao, Q. (2023, in press). Enhancing Museum Visiting Experience: Investigating the Relationships Between Augmented Reality Quality, Immersion, and TAM Using PLS-SEM. International Journal of Human–Computer Interaction, 1–12. https://doi.org/10.1080/10447318.2023.2227832

Chin, J., Diehl, V., & Norman, K. (1988). Development of an Instrument Measuring User Satisfaction of the Human-Computer Interface. Proceedings of the SIGCHI Conference on Human Factors in Computing Systems, 213–218. https://doi.org/10.1145/ 57167.57203

Condino, S., Carbone, M., Piazza, R., Ferrari, M., & Ferrari, V. (2020). Perceptual Limits of Optical See-Through Visors for Augmented Reality Guidance of Manual Tasks. *IEEE Transactions on Biomedical Engineering*, *67*(2), 411–419. https://doi.org/10.1109/ TBME.2019.2914517

Criollo, S., Abad, D., Martic, M., Velásquez, F., Pérez, J., & Luján, S. (2021). Towards a New Learning Experience through a Mobile Application with Augmented Reality in Engineering Education. *Applied Sciences*, *11*(11), 1–18. https://doi.org/10.3390/app1 1114921

Dalla, S., Di Nucci, D., Palomba, F., & Tamburri, D. (2020). Toward a Catalog of Software Quality Metrics for Infrastructure Code. *Journal of Systems and Software*, *170*, 1–8. https://doi.org/https://doi.org/10.1016/j.jss.2020.110726

Dargan, S., Bansal, S., Kumar, M., Mittal, A., & Kumar, K. (2023). Augmented Reality: A Comprehensive Review. *Archives of Computational Methods in Engineering*, *30*(2), 1057–1080. https://doi.org/10.1007/s11831-022-09831-7

Daşdemir, Y. (2022). Cognitive Investigation on the Effect of Augmented Reality-Based Reading on Emotion Classification Performance: A New Dataset. *Biomedical Signal Processing and Control*, *78*(1), 1–10. https://doi.org/https://doi.org/10.1016/j.bspc.2022.103942

Davis, F. (1989). Perceived Usefulness, Perceived Ease of Use, and User Acceptance of Information Technology. *MIS Quarterly*, *13*(3), 319–340. https://doi.org/10.2307/ 249008

Dede, C. (2009). Immersive Interfaces for Engagement and Learning. *Science*, *323*(5910), 66–69. https://doi.org/10.1126/science.11673

Di Serio, Á., Ibáñez, M., & Kloos, C. (2013). Impact of an Augmented Reality System on Students' Motivation for a Visual Art Course. *Computers & Education*, *68*(1), 586–596. https://doi.org/https://doi.org/10.1016/j.compedu.2012.03.002

Dumas, J., & Redish, J. (1993). *A Practical Guide to Usability Testing*. Greenwood Publishing Group Inc.

Dutta, R., Mantri, A., & Singh, G. (2022). Evaluating System Usability of Mobile Augmented Reality Application for Teaching Karnaugh-Maps. *Smart Learning Environments*, *9*(1), 6–27. https://doi.org/10.1186/s40561-022-00189-8

Edyanto, N., Ramli, S., Ibharim, N., Zahari, S., & Abdullah, M. (2021). Learn Idioms Using Augmented Reality. *International Journal of Multimedia and Recent Innovation (IJMARI)*, *3*(1), 11–16. https://doi.org/10.36079/lamintang.ijmari-0301.209

Fuvattanasilp, V., Fujimoto, Y., Plopski, A., Taketomi, T., Sandor, C., Kanbara, M., & Kato, H. (2021). SlidAR+: Gravity-Aware 3D Object Manipulation for Handheld Augmented Reality. *Computers & Graphics1*, *95*(1), 23–35. https://doi.org/https://doi.org/10.1016/j.cag.2021.01.005

Gao, M., Kortum, P., & Oswald, F. (2018). Psychometric Evaluation of the USE (Usefulness, Satisfaction, and Ease of use) Questionnaire for Reliability and Validity. *Proceedings of the Human Factors and Ergonomics Society Annual Meeting*, *62*(1), 1414–1418. https://doi.org/10.1177/1541931218621322

Georgiou, Y., & Kyza, E. (2017). The Development and Validation of the ARI Questionnaire: An Instrument for Measuring Immersion in Location-Based Augmented Reality Settings. *International Journal of Human-Computer Studies*, *98*, 24–37. https://doi.org/https://doi.org/10.1016/j.ijhcs.2016.09.014

Gonzalez, B., & Rosales, M. (2020). Measuring Student Motivation in a Statistics Course Supported by Podcast Using Reduced Instructional Materials Motivation Survey (RIMMS). Proceedings of the X International Conference on Virtual Campus (JICV), 1–4. https://doi.org/10.1109/JICV51605.2020.9375823

Greenslade, M., Clark, A., & Lukosch, S. (2022). User-Defined Interaction Using Everyday Objects for Augmented Reality First Person Action Games. Proceedings of the IEEE Conference on Virtual Reality and 3D User Interfaces Abstracts and Workshops (VRW), 842–843. https://doi.org/10.1109/VRW55335.2022.00272

Hart, S. (2006). Nasa-Task Load Index (NASA-TLX); 20 Years Later. *Proceedings of the Human Factors and Ergonomics Society Annual Meeting*, *50*(9), 904–908. https://doi.org/10.1177/154193120605000909

Hart, S., & Staveland, L. (1988). Development of NASA-TLX (Task Load Index): Results of Empirical and Theoretical Research. In Hancock, P., & Meshkati, N. (Eds.), *Human Mental Workload* (Vol. 52, pp. 139–183). North-Holland. https://doi.org/https://doi.org/10.1016/S0166-4115(08)62386-9

Hedenqvist, C., Romero, M., & Vinuesa, R. (2023). Improving the Learning of Mechanics Through Augmented Reality. *Technology, Knowledge and Learning*, *28*(1), 347–368. https://doi.org/10.1007/s10758-021-09542-1

Helin, K., Kuula, T., Vizzi, C., Karjalainen, J., & Vovk, A. (2018). User Experience of Augmented Reality System for Astronaut's Manual Work Support. *Frontiers in Robotics and AI*, *5*, 1–10. https://doi.org/10.3389/frobt.2018.00106

Hernández, L., López, J., Tovar, M., Vergara, O., & Cruz, V. (2021). Effects of Using Mobile Augmented Reality for Simple Interest Computation in a Financial Mathematics Course. *PeerJ Computer Science*, *7:e618*(1), 1–33.

Ibharim, N., Ramli, S., Zahari, S., Edyanto, N., & Abdullah, M. (2021). Learning History Using Augmented Reality. *International Journal of Multimedia and Recent Innovation (IJMARI)*, *3*(1), 1–10. https://doi.org/10.36079/lamintang.ijmari-0301.199

Iqbal, J., & Sidhu, M. S. (2022). Acceptance of Dance Training System Based on Augmented Reality and Technology Acceptance Model (TAM). *Virtual Reality*, *26*(1), 33–54. https://doi.org/10.1007/s10055-021-00529-y

Keenan, H., Duke, S., Wharrad, H., Doody, G., & Patel, R. (2022). Usability: An Introduction to and Literature Review of Usability Testing for Educational Resources in Radiation Oncology. *Technical Innovations & Patient Support in Radiation Oncology*, *24*(1), 67–72. https://doi.org/https://doi.org/10.1016/j.tipsro.2022.09.001

Keller, J. (2010a). *Motivational Design for Learning and Performance: The ARCS Model Approach* (1st ed.). Springer.

Keller, J. (2010b). Tools to Support Motivational Design. In Motivational Design for Learning and Performance: The ARCS Model Approach (pp. 267–295). Springer US. https://doi.org/10.1007/978-1-4419-1250-3_11

Kirakowski, J., & Corbett, M. (1993). SUMI: The Software Usability Measurement Inventory. *British Journal of Educational Technology*, *24*(3), 210–212. https://doi.org/https://doi.org/10.1111/j.1467-8535.1993.tb00076.x

Klemke, R., Limbu, B., Di Mitri, D., & Schneider, J. (2017). *Wearable Experience for Knowledge Intensive Training*, Technical Report: Horizon 2020, 1–30. https://10.3030/687669

Korkut, S., Mele, E., & Cantoni, L. (2022). User Experience and Usability: The Case of Augmented Reality. In Xiang, Z., Fuchs, M., Gretzel, U., & Höpken, W. (Eds.), *Handbook of e-Tourism* (pp. 1017–1038). Springer International Publishing. https://doi.org/10.1007/978-3-030-48652-5_62

Koutromanos, G., & Kazakou, G. (2023). Augmented Reality Smart Glasses Use and Acceptance: A Literature Review. *Computers & Education: X Reality*, *2*, 1–14. https://doi.org/https://doi.org/10.1016/j.cexr.2023.100028

Lewis, J. (1995). IBM Computer Usability Satisfaction Questionnaires: Psychometric Evaluation and Instructions for Use. *International Journal of Human–Computer Interaction*, *7*(1), 57–78. https://doi.org/10.1080/10447319509526110

Lewis, J. (2018). Measuring Perceived Usability: The CSUQ, SUS, and UMUX. *International Journal of Human–Computer Interaction*, *34*(12), 1148–1156. https://doi.org/10.1080/10447318.2017.1418805

Li, K., & Keller, J. (2018). Use of the ARCS Model in Education: A Literature Review. *Computers & Education*, *122*, 54–62. https://doi.org/https://doi.org/10.1016/j.compedu.2018.03.019

Lin, P., Chai, C., Jong, M., Dai, Y., Guo, Y., & Qin, J. (2021). Modeling the Structural Relationship Among Primary Students' Motivation to Learn Artificial Intelligence. *Computers and Education: Artificial Intelligence*, *2*, 1–7. https://doi.org/https://doi.org/10.1016/j.caeai.2020.100006

Loorbach, N., Peters, O., Karreman, J., & Steehouder, M. (2014). Validation of the Instructional Materials Motivation Survey (IMMS) in a Self-Directed Instructional Setting Aimed at Working with Technology. *British Journal of Educational Technology (BJET)*, *46*(1), 204–218. https://doi.org/10.1111/bjet.12138

Lund, A. (2001). Measuring Usability with the USE Questionnaire. *Usability Interface*, *8*(2), 3–6.

Maier, P., Tönnis, M., & Klinker, G. (2009). Augmented Reality for Teaching Spatial Relations. *Proceedings of the Conference of the International Journal of Arts & Sciences*, 1–8.

Marangunić, N., & Granić, A. (2015). Technology Acceptance Model: A Literature Review from 1986 to 2013. *Universal Access in the Information Society*, *14*(1), 81–95. https://doi.org/10.1007/s10209-014-0348-1

Marsden, E., & Torgerson, C. J. (2012). Single Group, Pre- and Post-test Research Designs: Some Methodological Concerns. *Oxford Review of Education*, *38*(5), 583–616. https://doi.org/10.1080/03054985.2012.731208

Marshall, G. (2005). The Purpose, Design and Administration of a Questionnaire for Data Collection. *Radiography*, *11*(2), 131–136. https://doi.org/https://doi.org/10.1016/j.radi.2004.09.002

Meister, P., Wang, K., Dorneich, M., Winer, E., Brown, L., & Whitehurst, G. (2023). Evaluation of Augmented Reality Interactive Print for General Aviation Weather Training. *Journal of Air Transportation*, *2023*(1), 1–10. https://doi.org/10.2514/1.D0364

Miranda, E., Vergara, O., Cruz, V., García, J., & Favela, J. (2016). Study on Mobile Augmented Reality Adoption for Mayo Language Learning. *Mobile Information Systems*, *2016*, 1–15. https://doi.org/10.1155/2016/1069581

Nazar, M., Aisyi, R., Rahmayani, R., Hanum, L., Rusman, R., Puspita, K., & Hidayat, M. (2020). Development of Augmented Reality Application for Learning the Concept of Molecular Geometry. *Journal of Physics: Conference Series*, *1460*(1), 12083. https://doi.org/10.1088/1742-6596/1460/1/012083

Olsson, T. (2013). Concepts and Subjective Measures for Evaluating User Experience of Mobile Augmented Reality Services. In Huang, W., Alem, L., & Livingston, M. (Eds.), *Human Factors in Augmented Reality Environments* (pp. 203–232). Springer New York. https://doi.org/10.1007/978-1-4614-4205-9_9

Otte, C., Bolling, M., Stevenson, M., Ejbyet, N., Nielsen, G., & Bentsen, P. (2019). Education Outside the Classroom Increases Children's Reading Performance: Results from a One-Year Quasi-Experimental Study. *International Journal of Educational Research*, *94*, 42–51. https://doi.org/https://doi.org/10.1016/j.ijer.2019.01.009

Oyman, M., Bal, D., & Ozer, S. (2022). Extending the Technology Acceptance Model to Explain How Perceived Augmented Reality Affects Consumers' Perceptions. *Computers in Human Behavior*, *128*, 1–10. https://doi.org/https://doi.org/10.1016/j.chb.2021.107127

Polvi, J., Taketomi, T., Yamamoto, G., Dey, A., Sandor, C., & Kato, H. (2016). SlidAR: A 3D Positioning Method for SLAM-Based Handheld Augmented Reality. *Computers & Graphics*, *55*(1), 33–43. https://doi.org/https://doi.org/10.1016/j.cag.2015.10.013

Pratama, A., & Sukirman. (2023). Development of Augmented Reality Multiple Markers Application Used for Interactive Learning Media. *Sinkron: Jurnal Dan Penelitian Teknik Informatika*, *1*(3), 1326–1334.

Ramli, S., Zahari, S., Edyanto, N., Abdullah, M., & Ibharim, N. (2021). Travel to Southeast Asia: Learning About Southeast Asia through Augmented Reality. *International Journal of Multimedia and Recent Innovation (IJMARI)*, *3*(2), 17–28. https://doi.org/10.36079/lamintang.ijmari-0302.272

Santos, M., Lübke, A., Taketomi, T., Yamamoto, G., Rodrigo, M., Sandor, C., & Kato, H. (2016). Augmented Reality as Multimedia: The Case for Situated Vocabulary Learning. *Research and Practice in Technology Enhanced Learning*, *11*(4), 1–23. https://doi.org/10.1186/s41039-016-0028-2

Santos, M., Polvi, J., Taketomi, T., Yamamoto, G., Sandor, C., & Kato, H. (2015). Toward Standard Usability Questionnaires for Handheld Augmented Reality. *IEEE Computer Graphics and Applications*, *35*(5), 66–75. https://doi.org/10.1109/MCG.2015.94

Santos, M., Taketomi, T., Sandor, C., Polvi, J., Yamamoto, G., & Kato, H. (2014). A Usability Scale for Handheld Augmented Reality. Proceedings of the 20th ACM Symposium on

Virtual Reality Software and Technology (VRST), 167–176. https://doi.org/10.1145/2671015.2671019

Shin, D. (2019). How does Immersion Work in Augmented Reality Games? A user-Centric View of Immersion and Engagement. *Information, Communication & Society*, 22(9), 1212–1229. https://doi.org/10.1080/1369118X.2017.1411519

Slaughter, L., Norman, K., & Shneiderman, B. (1995). Assessing Users' Subjective Satisfaction with the Information System for Youth Services (ISYS). *VA Tech Proceedings of Third Annual Mid-Atlantic Human Factors Conference*, 164–170.

Ssemugabi, S., & De Villiers, R. (2007). A Comparative Study of Two Usability Evaluation Methods Using a Web-Based E-Learning Application. *Proceedings of the 2007 Annual Research Conference of the South African Institute of Computer Scientists and Information Technologists on IT Research in Developing Countries*, 132–142.

Thabit, A., Benmahdjoub, M., van Veelen, M., Niessen, W., Wolvius, E., & van Walsum, T. (2022). Augmented Reality Navigation for Minimally Invasive Craniosynostosis Surgery: A Phantom Study. *International Journal of Computer Assisted Radiology and Surgery*, 17(8), 1453–1460. https://doi.org/10.1007/s11548-022-02634-y

Tomara, M., & Gouscos, D. (2019). A Case Study: Visualizing Coulomb Forces with the Aid of Augmented Reality. *Journal of Educational Computing Research*, 57(7), 1626–1642. https://doi.org/10.1177/0735633119854023

Tzortzoglou, F., & Sofos, A. (2023). Evaluating the Usability of Mobile-Based Augmented Reality Applications for Education: A Systematic Review. In Bratitsis, T., (Ed.), Research on E-Learning and ICT in Education: Technological, Pedagogical, and Instructional Perspectives (pp. 105–135). Springer International Publishing. https://doi.org/10.1007/978-3-031-34291-2_7

Uriarte, A., Ibáñez, M., Zatáraín, R., & Barrón, M. (2022). Higher Immersive Profiles Improve Learning Outcomes in Augmented Reality Learning Environments. *Information*, 13(5), 1–11. https://doi.org/10.3390/info13050218

van Riesen, S., Gijlers, H., Anjewierden, A., & de Jong, T. (2022). The Influence of Prior Knowledge on the Effectiveness of Guided Experiment Design. *Interactive Learning Environments*, 30(1), 17–33. https://doi.org/10.1080/10494820.2019.1631193

Villagran, D., Luviano, D., Pérez, L., Méndez, L., & Garcia, F. (2023). Applications Analyses, Challenges and Development of Augmented Reality in Education, Industry, Marketing, Medicine, and Entertainment. *Applied Sciences*, 13(5), 1–30. https://doi.org/10.3390/app13052766

Vlachogianni, P., & Tselios, N. (2022). Perceived Usability Evaluation of Educational Technology Using the System Usability Scale (SUS): A Systematic Review. *Journal of Research on Technology in Education*, 54(3), 392–409. https://doi.org/10.1080/15391523.2020.1867938

Wang, C., Hsiao, C., Tai, A., & Wang, M. (2023). Usability Evaluation of Augmented Reality Visualizations on an Optical See-Through Head-Mounted Display for Assisting Machine Operations. *Applied Ergonomics*, 113, 1–10. https://doi.org/https://doi.org/10.1016/j.apergo.2023.104112

Xue, H., Sharma, P., & Wild, F. (2019). User Satisfaction in Augmented Reality-Based Training Using Microsoft HoloLens. *Computers*, 8(1), 1–23. https://doi.org/10.3390/computers8010009

Yilmaz, R., Topu, F., & Takkaç, A. (2022). An Examination of Vocabulary Learning and Retention Levels of Pre-School Children Using Augmented Reality Technology in English Language Learning. *Education and Information Technologies*, 27(5), 6989–7017. https://doi.org/10.1007/s10639-022-10916-w

Zhang, B., & Robb, N. (2021). Immersion Experiences in a Tablet-Based Markerless Augmented Reality Working Memory Game: Randomized Controlled Trial and User Experience Study. *JMIR Serious Games*, *9*(4), 1–14. https://doi.org/10.2196/27036

Zigart, T., & Schlund, S. (2020). Evaluation of Augmented Reality Technologies in Manufacturing—A Literature Review. In Nunes, I. (Ed.), *Advances in Human Factors and Systems Interaction* (pp. 75–82). Springer International Publishing.

7 Case Study I
ARGEO: Augmented Reality Prototype for Learning Geography

7.1 INTRODUCTION

Geography is a science that mixes natural and social sciences to understand how the world works. Geography presents the study of the Earth, focusing on the terrestrial relief, its distribution, location, and all the components that develop on it. Moreover, geography must be learned in primary and secondary education because it is considered a relevant and essential science for the academic formation of young people (Bunge, 1973).

Despite how important learning geography is, students often consider it a useless and boring science. Frequently, geography teaching focuses only on memorizing places or countries' names and capitals. Another problem is that the concepts of geography cannot be experienced in laboratories. In addition, it is typical for school geographers to be more researchers than teachers. Consequently, they do not use teaching methods based on the abilities and capabilities of the students. In addition, the challenge of teaching and learning geography has increased due to the COVID-19 pandemic (Schultz & DeMers, 2020).

According to Piotrowska et al. (2019), the main problem is that many professors who teach geography are historians, many of whom are recognized professionals but are not geographers. In short, history is often confused with geography, despite them being different. Therefore, today, the search continues to improve the teaching–learning strategies used in geography education.

One of the strategies that can be used to solve the problem of lack of interest in the study of geography is to incorporate information and communication technologies (ICTs) into the didactic contents. ICTs have been considered a way to enhance learning in many sciences, including geography. Interested readers are encouraged to review the work by Chang and Sheng (2018) to gain insight into the use of technology in teaching geography.

AR is one of the ICTs that can support students in educational settings. Moreover, many studies have been presented in the literature in which the benefits of using AR for teaching geography have been documented (Adedokun-Shittu et al., 2020; Shelton & Hedley, 2002; Turan et al., 2018; Volioti et al., 2022). Most studies have recommended designing AR prototypes for teaching geography, considering game-based design principles aiming to promote students' interaction and collaboration.

After identifying the need to include technologies and diversify the strategies for teaching geography, this chapter proposes the construction of an AR prototype named "ARGeo" to support fifth-grade Mexican students in learning geography. The work results from the Bs.C. computer science dissertation by Coronado (2020).

The prototype comprises six topics aimed at learning about the natural components of the Earth. It also includes 3D models that contain texts and animations with which the user can interact. Motivation, technology acceptance, immersion, quality, and student achievement were measured.

The rest of the chapter is organized as follows. Section 7.2 presents the methodology followed to build the AR prototype. The experiments and results are shown in Section 7.3. Finally, the conclusions are presented in Section 7.4.

7.2 ARGEO DESIGN

Several meetings were held to discuss the software model on which the prototype should be based. Finally, the V-model for the software development life cycle was selected to describe the relationships of each part the prototype will include. The V-model is an extension of the traditional waterfall model and is also known as the verification and validation model. All the processes are executed sequentially following a V shape in the V-model (Hon & Feng, 2023).

The V model is based on the association of a testing stage for each development stage. Therefore, a corresponding testing phase must be conducted for every stage in the development cycle. The verification phases are located on one side of the V, and the validation phases are on the other side. The verification includes four stages: (i) requirements gathering and analysis, (ii) system design, (iii) architectural design, and (iv) module design. The validation includes four stages: (i) unit testing for testing the module design, (ii) integration testing for testing the architectural design, (iii) system testing for testing the system design, and (iv) acceptance testing for testing the requirements gathering and analysis. The coding stage is located at the bottom of the V shape (Jabangwe et al., 2018).

The V model was used because tests are performed at each stage, which allows early detection of errors. In case of finding an error, a redesign of the component is conducted until the expected result is obtained.

The project began by searching for a school institution that required support from a technological learning tool. The search detected a public elementary school in Ciudad Juarez, Mexico. After several meetings with the school's director, the conclusion was reached that the children in the fifth year presented problems in learning the subjects of geography and history. However, the teachers selected the subject of geography as the one that urgently needs to include new teaching resources.

According to the professors, the requirements that the prototype must include are the following:

- Show the information in an attractive way for users.
- Serve as support to reinforce student learning.
- Arouse the interest of students.

- Allow the student to learn at his/her own pace.
- Access the topics in an organized and individual way.

After the analysis, it was concluded that AR could help to comply with the requirements. Therefore, the following functional requirements were defined: (a) AR will be activated utilizing markers, (b) each marker will have an associated model that will be superimposed on the video scene, (c) each scene must show interactive elements related to a learning topic, (d) the prototype must contain activities to reinforce what has been learned, and (e) each activity will award points that can be exchanged for prizes.

In Mexico, the teaching of geography for fifth-year children includes five thematic axes: (i) geographic space and maps, (ii) natural components of the Earth, (iii) social and cultural components, (iv) economic components, and (v) quality of life, environment, and the prevention of disasters (Acosta et al., 2019).

As a result of the consensus with the teachers of the subject, the second thematic axis about the natural components of the Earth was selected to include in the AR prototype. The sub-themes included in the thematic axis are the following:

1. The layers of the Earth.
2. Sliding movement.
3. Separation movement.
4. Convergence movement (continental–continental).
5. Convergence movement (oceanic–continental).
6. Earthquakes.

The prototype runs on Android devices, and Unity and Vuforia were employed for programming. Blender was employed to design all the 3D models. The final graphical user interface of the prototype is shown in Figure 7.1. The first button shows the instructions to interact with AR content. The second button must be pressed to experiment with AR. Once the topic is selected, the corresponding markers must be

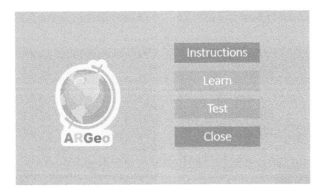

FIGURE 7.1   The main screen of the ARGeo prototype.

displayed to perform the learning. The third button shows a test with multiple-choice questions. Once the test is finished, the score appears according to the number of correct answers. Depending on the score, the user wins points that can be exchanged for rewards. Finally, the fourth button closes the application.

7.2.1 VIRTUAL MODELS DESIGN

The objective of the prototype is to help the learning of geography through textual and visual elements. Therefore, all the models were selected from the geography book of the fifth year of primary school (Acosta et al., 2019). Each topic has a 3D model and a terrain associated with it in order to support learning. Presenting a virtual scenario helps to have a better perspective of what is being explained. Therefore, the student can see the explanation visually. Also, informative text of what is being observed is displayed.

A 3D model allows one to view an object referenced by a context from different angles. Also, the student can interact with the 3D model to understand how the process is realized from different perspectives. All the 3D models were generated using the Blender open-source software (Soni et al., 2023). Moreover, all the models were based on a polygon aesthetic design, including terrains, spheres, cubes, and planes. Six models with terrains were created. Figure 7.2 shows some of the models and terrains designed.

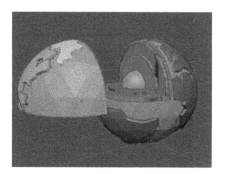
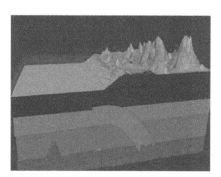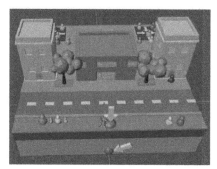

FIGURE 7.2 Examples of 3D models and terrains designed for ARGeo.

Blender was also used to generate the animation of the virtual models. Thanks to the animations, 3D models move and interact with each other, aiming to give details about the topics to be learned. The rigging technique was the basis for the object animation. Rigging is a technique employed to represent a 3D character model using a series of interconnected digital bones and to determine the range of movement of the object (Pan et al., 2009).

7.2.2 MARKERS DESIGN

Once all the 3D models were finished, the markers were designed and used to activate the AR. The Brosvision marker generator was employed to create the background of the markers (Brosvision s.r.o., 2023). The markers were generated in full color, including lines, triangles, and quadrilaterals. A total of six markers were generated, one for each topic. An image of a 3D model was added to each of the markers generated with Brosvision to make them more attractive to users. The insertion of the images facilitates the identification of each marker. The image insertion was conducted with the GIMP software (Whitt, 2023).

Each image included a descriptive text concerning the six topics addressed for better user discrimination. Figure 7.3 shows the six markers generated for the ARGeo prototype.

7.2.3 AR SCENES

A new window containing three buttons is displayed when the "learn" button is pressed in the prototype. The six topics were grouped into three buttons. The first button must be pressed to learn the topic of the layers of the Earth. The second button

FIGURE 7.3   The six markers employed in ARGeo.

is used to learn about the movements of the tectonic plates. The third button shows the theme of earthquakes. Figure 7.4 shows the screen for topic selection.

When the button regarding tectonic plates is pressed, a new screen with three buttons is displayed. Figure 7.5 shows the screen related to tectonic plate movements. The first button must be pressed to learn about convergency movements. A new screen, including the continental and oceanic buttons, is displayed when this button is pressed, as shown in Figure 7.6. The second one must be pressed to learn about sliding movements. Finally, the third button must be pressed to learn about separation movements.

If the user presses the button test, a new screen with three buttons is displayed, as shown in Figure 7.7. The instruction button displays the explanations to conduct the

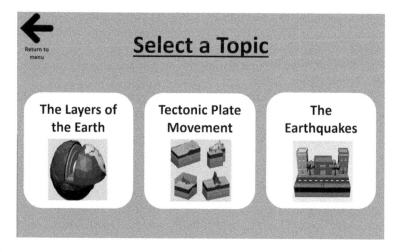

FIGURE 7.4 The screen for main topics selection.

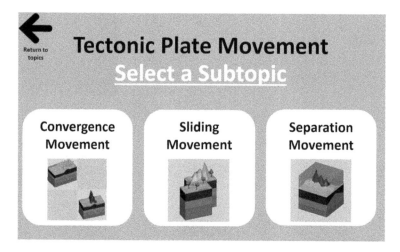

FIGURE 7.5 The screen to learn about tectonic plate movement.

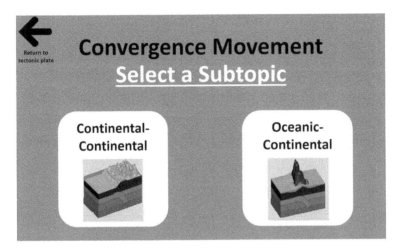

FIGURE 7.6 The screen about oceanic and continental movements.

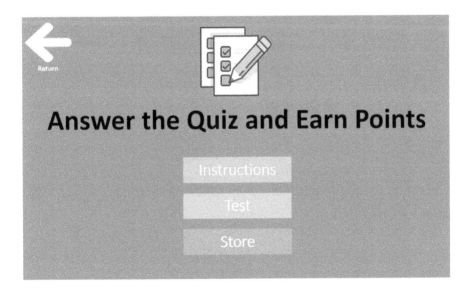

FIGURE 7.7 The screen for tests.

tests. The button tests open the screen to start the test. Twelve random questions are included in each test. For each correct answer, 50 points are assigned. Therefore, a maximum of 600 points can be earned.

Figure 7.8 shows an example of the questions included in the tests. The button store displays the screen to exchange the points for rewards, as shown in Figure 7.9.

The user must select and show the marker associated with each topic to be learned. Then, a new screen is opened, requesting to show the marker. The AR scene is drawn when a marker is recognized. Therefore, the 3D model is inserted into the video

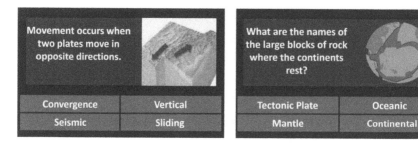

FIGURE 7.8 Examples of test questions.

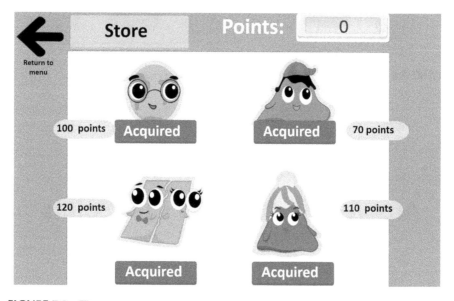

FIGURE 7.9 The store screen.

stream. The user can push the button animation to observe the movements of the virtual objects. The button information creates virtual texts overlaid in the video stream to explain the observed process. Figure 7.10 shows an example of the AR scenes included in ARGeo.

7.2.4 DATA COLLECTION INSTRUMENTS

A set of surveys was created to assess students' motivation, technology usage with the technology acceptance model (TAM), and prototype quality when using ARGeo. Also, a pool with 60 practice exercises was created to assess students' achievement. In the following, a brief explanation of the instruments employed is offered. The details for each instrument can be found in Chapter 6.

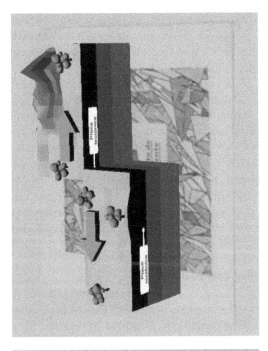

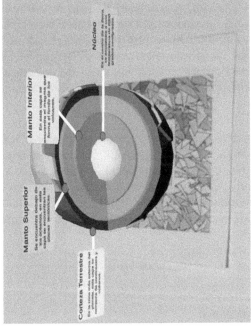

FIGURE 7.10 AR scenes included in ARGeo.

(A) Motivation

A teacher's main challenge is encouraging students to participate, be interested, and be motivated to learn. One of the tools that can help achieve these aspects is motivational design. Motivation affects what, how, and when the learners learn, and it is directly related to the development of students' attitudes and persistent efforts toward achieving a goal (Lin et al., 2021).

Keller's Attention, Relevance, Confidence, and Satisfaction (ARCS) model provides guidelines for designing and developing strategies to motivate students' learning (Li & Keller, 2018). The instructional materials motivation survey (IMMS) assesses students' motivation based on the ARCS model and includes 36 items distributed as 12 items for (A), nine items for (R), nine items for (C), and six items for (S). The problem is that IMMS is long, and not all items are necessary.

Therefore, the Reduced Instructional Materials Motivation Survey (RIMMS) assessed students' motivation. RIMMS comprises 12 five-point Likert scale items, three for each ARCS dimension. The RIMMS original version was translated and adapted to the geography lesson. The minimum score on the RIMMS survey is 12, and the maximum is 60, with a midpoint of 36 (Loorbach et al., 2014).

(B) Augmented Reality Immersion

Immersion means being completely immersed in an artificial world. AR's goal is that, at some point, the user will be unable to distinguish the real from the virtual. The more communication between the human senses and the AR system, the more immersive the system will be. Therefore, it can be desirable to know if the AR system performs well or better than others in learning geography (Liberatore & Wagner, 2021).

The Augmented Reality Immersion (ARI) questionnaire was employed to measure students' immersion using the ARGeo prototype (Georgiou & Kyza, 2017). ARI comprises 21 seven-point Likert scale items regarding engagement, engrossment, and total immersion. The ARI original version was translated and adapted to the geography lesson. The minimum score on the survey is 21, and the maximum is 147.

(C) Quality

Software quality describes the desirable characteristics of a software product. Attributes such as design, usability, operability, security, compatibility, maintainability, and functionality can be considered to define software quality (Dalla et al., 2020). However, questions such as how fast the system responds, how difficult it is to manipulate the system and markers, and to what extent the illumination affects marker recognition must be answered to measure the quality of an AR prototype.

The quality survey was designed based on the work by Barraza et al. (2015). The survey comprises 10 five-point Likert scale items. The minimum score on the survey is 10, and the maximum is 50.

(D) Technology Acceptance

The Technology Acceptance Model (TAM) was developed by (Davis, 1989) to explain how to encourage users to accept new technologies. TAM suggested that the

Perceived Ease of Use (PEU) and the Perceived Usefulness (PU) are determinants to explain what causes the Intention of a person To Use (ITU) a technology.

TAM survey comprises 11 five-point Likert scale items. The minimum score on the survey is 11, and the maximum is 55.

(E) Student Achievement

Academic achievement is the extent to which a student has accomplished specific goals that focus on activities in instructional environments (Bernacki et al., 2020). Student achievement was related to how capable the students were when solving questions regarding the natural components of the Earth.

The professors helped to create a pool of 60 questions about the themes learned with ARGeo, 10 for each natural component of the Earth subtopic. The questions were divided into two groups. Hence, 30 questions were employed for the pre-test and 30 for the post-test. A total of 12 questions were randomly selected for each test. Moreover, the same complexity was ensured for both tests.

7.3   EXPERIMENTS AND RESULTS

The prototype was installed in three mobile devices (Motorola, Samsung, and Huawei) and two tablets (Samsung and Lenovo) to verify the correct execution and visualization of 3D models and descriptive texts in different architectures. One experienced software developer tested the prototype and recommended modifications related to the objects' locations, colors, and sizes. After making the suggested changes, the test with fifth-grade students was conducted.

The tests were conducted in two sessions in November 2019, with one day of difference, each lasting one hour and a half. A classroom at a public school in Ciudad Juarez, Mexico, was used as an educational setting. Two professors helped to conduct the experiments. Twenty-five Mexican students from fifth grade participated in the study. Of the 25 participants, $n = 9$ (36%) were female, and $n = 16$ (64%) were male. The participants' ages ranged from 10 to 12 years.

In the first session, the first professor explained the topics of the natural components of the Earth using traditional tools like a whiteboard, marker, and printed book. Students were then asked to take part in a pre-test consisting of 12 test exercises and fill out the survey to determine their motivation for learning with traditional materials. Both instruments were delivered to the students in printed form.

At the end of the first session, students were requested to get an Android-based mobile device for the next day to experiment with AR. Fortunately, most students brought an Android mobile device. Children who could not get a device were loaned one. The ARGeo prototype was installed on all devices the day before the second session. Hence, mobile devices (smartphones and tablets) with different features were used.

A video explaining how to use the system was created. On the day of testing the ARGeo prototype (second session), the second teacher explained the test's purpose to the students and showed them how to use the system. Each child received a set of printed markers. The average time to interact with the prototype was half an hour.

Then, students were asked to complete the post-test consisting of 12 test exercises and fill out the surveys regarding motivation, immersion, technology acceptance, and quality. All the instruments were delivered to the students in a printed form.

Data collected from the surveys was captured to create a database with IBM SPSS software (Sen & Yildirim, 2022). After capturing the data, it was verified that all the information was congruent. Fortunately, no errors were found. The internal reliability of the surveys was measured with Cronbach's alpha (α) following the protocol explained by Hernández et al. (2021). Therefore, the items for all surveys were considered valid.

7.3.1 Students' Motivation

The study allowed us to assess whether a significant difference in motivation is obtained when comparing the professor's lesson and ARGeo. The 25 students filled out the surveys regarding motivation in the pre-test and post-test sessions. The results obtained regarding mean (*M*) and standard deviation (*SD*) for each item are shown in Table 7.1.

As observed, all scores exceeded the central value of the scale. The AR prototype received the best evaluation for all the categories compared with traditional learning. However, the evaluations for traditional learning are also good, meaning that teachers use effective techniques that motivate students. The mean for the pre-test was 3.88, and for the post-test was 4.00. Therefore, a difference of 0.12 was obtained. The difference is statistically significant, representing a motivation increase of 2.4%.

7.3.2 Augmented Reality Immersion (ARI)

The ARI questionnaire was employed to measure students' immersion on the levels of engagement, engrossment, and total immersion, and subscales of interest, usability, emotional attachment, focus of attention, presence, and flow. The results obtained regarding mean (*M*) and standard deviation (*SD*) for each item are shown in Table 7.2.

As observed, all scores exceeded the central value of the seven-point Likert scale. Usability obtained the highest score, and emotional attachment was the lowest. By looking at the results, it can be concluded that students showed interest when interacting with the prototype. The usability measure shows that most students interact without difficulty with the prototype. Regarding the emotional attachment and focus of attention, it was observed that students were curious and excited to perform the activity. In addition, even when the students were distracted, they were interested in continuing the activity.

The students expressed that, at some point, they felt the interaction with the virtual objects was real. Despite the duration of the activity, the students remained interested in doing it. Therefore, considering a total mean of 5.28, it can be concluded that the immersion offered by ARGeo is suitable for young students.

TABLE 7.1
ARCS Results for the Pre-test and Post-test

General Data					
Name (s):			**Surname:**		
Age:					
Gender:	o (Male)		o (Female)		
ARCS Professor	**Mean**	**SD**	**ARCS ARGeo**	**Mean**	**SD**
Attention (A)	**3.80**	**1.00**	**Attention (A)**	**3.86**	**1.10**
A1. The quality of the materials used helped to hold my attention.	3.58	1.44	A1. The quality of the contents displayed helped to hold my attention.	4.00	1.08
A2. The way the information was organized helped keep my attention.	3.83	0.55	A2. The way the information was organized (buttons, menus) helped keep my attention.	3.75	1.08
A3. The various readings, exercises, and illustrations helped keep my attention on the explanations.	4.00	1.00	A3. The various 2D models and interactions helped keep my attention on the explanations.	3.83	1.14
Relevance (R)	**3.80**	**1.02**	**Relevance (R)**	**3.89**	**1.08**
R1. It is clear to me how the content of this lesson is related to things I already know.	3.33	1.17	R1. It is clear to me how the content of ARGeo is related to things I already know.	3.35	1.36
R2. The content and style of explanations convey the impression that working with the earth's natural components is worth it.	4.58	0.64	R2. The content and style of explanations used by ARGeo convey the impression that working with the natural components of the earth is worth it.	4.17	0.89

(continued)

TABLE 7.1 (Continued)
ARCS Results for the Pre-test and Post-test

R3. The content of this lesson will be useful to me.	3.5	1.25	R3. The content of ARGeo will be useful to me.	4.17	0.98
Confidence (C)	**3.94**	**1.21**	**Confidence (C)**	**4.11**	**0.94**
C1. As I worked on this lesson, I was confident I could learn about the natural components of the earth well.	4.17	1.34	C1. As I worked with ARGeo, I was confident that I could learn about the natural components of the earth well.	4.33	0.84
C2. After working with this lesson for a while, I was confident that I could pass a test about the natural components of the earth.	3.92	0.86	C2. After working with ARGeo for a while, I was confident that I would be able to pass a test about the natural components of the earth.	4.25	0.82
C3. The good organization of the content helped me be confident that I would learn about the natural components of the earth.	3.75	1.42	C3. The good organization of ARGeo helped me be confident that I would learn about the natural components of the earth.	3.75	1.16
Satisfaction (S)	**4.00**	**1.22**	**Satisfaction (S)**	**4.16**	**0.91**
S1. I enjoyed working with this lesson so much that I was stimulated to keep working.	3.75	1.36	S1. I enjoyed working with ARGeo so much that I was stimulated to keep working.	4.00	0.91
S2. I really enjoyed working with this geography lesson.	4.08	1.25	S2. I really enjoyed working with ARGeo.	4.25	0.82
S3. It was a pleasure to work with such a well-designed lesson.	4.17	1.06	S3. It was a pleasure to work with such a well-designed prototype.	4.25	1.01

TABLE 7.2
Results Obtained with ARI

Immersion Level	Scale	Mean	SD
Engagement	**Interest**		
	I1. I liked the activity because it was novel.	5.92	1.11
	I2. I liked the type of activity.	5.67	1.59
	I3. I wanted to spend the time to complete the activity successfully.	5.83	1.21
	I4. I wanted to spend time participating in the activity.	5.83	1.21
	Usability		
	U1. It was easy for me to use the AR application.	6.17	1.14
	U2. I found the AR application confusing.	4.08	1.75
	U3. The AR application was unnecessarily complex.	3.58	1.11
	U4. I did not have difficulties in controlling the AR application.	4.92	2.17
Engrossment	**Emotional Attachment**		
	EA1. I was curious about how the activity would progress.	5.42	1.32
	EA2. I was often excited since I felt part of the activity.	5.75	1.36
	EA3. I often felt suspense by the activity.	4.33	1.59
	Focus of attention		
	FA1. If interrupted, I looked forward to returning to the activity.	5	1.52
	FA2. Everyday thoughts and concerns faded out during the activity.	5.33	1.1
	FA3. I was more focused on the activity rather than on any external distractions.	5.83	1.34
Total Immersion	**Presence**		
	P1. The activity felt so authentic that it made me think that the virtual characters/objects existed for real.	5.08	2.13
	P2. I felt that what I was experiencing was something real instead of a fictional activity.	5.58	1.38
	P3. I was so involved in the activity that I sometimes wanted to interact directly with the virtual characters/objects.	5.25	1.29
	P4. I was so involved that I felt that my actions could affect the activity.	4.75	1.87
	Flow		
	F1. I did not have any irrelevant thoughts or external distractions during the activity.	5.58	1.32
	F2. The activity became the unique and only thought occupying my mind.	5.83	1.34
	F3. I lost track of time as if everything just stopped, and the only thing I could think about was the activity.	5.25	1.42

TABLE 7.3
Results Obtained from Quality Measurement

Quality Questions	Mean	SD
Q1. ARGeo showed all the concepts explained by the teacher.	4.5	0.87
Q2. The results obtained with ARGeo were correct.	4.5	0.5
Q3. The colors used for conversions were adequate.	5.00	0.0
Q4. The texts and numbers displayed by ARGeo were legible.	4.5	0.5
Q5. The size of the buttons allowed the easy manipulation of ARGeo.	4.75	0.43
Q6. ARGeo velocity of response to carry out the calculations was fast.	4.5	0.5
Q7. The classroom illumination was adequate.	4.75	0.43
Q8. The manipulation of the electronic device I used was straightforward.	4.5	0.5
Q9. Markers' manipulation was easy.	4.5	0.5
Q10. The manipulation of the device in conjunction with the markers was easy.	4.75	0.43

7.3.3 QUALITY

A survey was conducted to gain insights about ARGEo quality. The survey contained 10 items, and as can be observed in Table 7.3, all the measures are equal to or greater than 4.5. The mean value of 4.62 demonstrates that students consider ARGeo a good-quality prototype.

The students considered that the colors used were the best characteristic of the prototype. The lighting in the room, the ease of handling the prototype, and the size of the buttons also received good comments. The minimum values were obtained for concept explanations, results, the speed of response, mobile device manipulation, and markers manipulation.

Data obtained from the quality survey follow a normal distribution. According to Barraza et al. (2015), quality results greater than 4.2 are considered excellent. Moreover, quality influences the students' intention to use ARGeo.

7.3.4 TECHNOLOGY ACCEPTANCE MODEL

It is often mentioned that AR prototypes are helpful when used in educational settings. However, knowing if students are willing to use technology daily is appropriate. The TAM was employed to measure the students' acceptance of ARGeo. The students answered the TAM survey containing 11 questions. Table 7.4 shows the results obtained for each TAM item.

As observed, all the items exceeded the central value of the five-point Likert scale. According to the results, students expressed their intention to use ARGeo.

TABLE 7.4
Results Obtained from TAM

TAM	Mean	SD
Perceived Usefulness (PU)	**4.62**	**0.53**
PU1. I could improve my learning performance by using ARGeo.	4.5	0.87
PU2. I could enhance my simple interest proficiency by using ARGeo.	5.00	0.00
PU3. I think ARGeo is useful for learning purposes.	4.75	0.43
PU4. Using ARGeo, it will be easy to remember the concepts related to calculating simple interest.	4.25	0.83
Perceived Ease of Use (PEU)	**4.75**	**0.28**
PEU1. I think ARGeo is attractive and easy to use	5.00	0.00
PEU2. Learning to use ARGeo was not a problem due to my familiarity with the technology used.	4.5	0.5
PEU3. The marker detection was fast.	4.75	0.43
PEU4. The tasks related to the manipulation of controls were simple to execute.	4.5	0.5
PEU5. I was able to locate the areas for conversions and calculations quickly.	5	0
Intention to Use SICMAR (ITU)	**4.62**	**0.41**
ITU1. I want to use the app in the future if I have the opportunity.	4.25	0.83
ITU2. The main concepts of ARGeo can be used to learn other topics.	5	0

7.3.5 Assessment of Students' Achievement in Practice Tests

Both professors reviewed and validated the students' responses for practice tests. Each test includes 12 items (two for each subtopic), and each correct answer totals 50 points. Therefore, the grades ranged from 0 to 600. The final grade is obtained employing a rule of three, where 600 equals the maximum grade of 10.

Table 7.5 summarizes the results obtained from the tests. Fifty questions were answered for each topic. Therefore, 300 questions were evaluated for the pre-test and 300 for the post-test. The mean grade for the pre-test was 7.83, while the post-test was 8.53. Hence, an increase of 7% was observed in post-test grades compared with the pre-test. Moreover, the more complex topic was the movement of convergence (oceanic–continental) because, in both tests, students made more mistakes. On the contrary, the movement of convergence (continental–continental) obtained better grades.

Students obtained an average grade of 8.53 with ARGeo and 7.83 with the professor's material. The difference of 0.7 is statistically significant. Therefore, we can conclude that learning with ARGeo achieves higher scores in natural components of the Earth tests than students exposed to traditional learning.

These results confirm that using AR in the classroom benefits students. The use of ARGeo increases students' motivation when learning about geography. Professors

TABLE 7.5
Results for the Practice Tests

Topic tested	Pre-test		Post-Test	
	Correct	Incorrect	Correct	Incorrect
The layers of the earth	39	11	42	8
Sliding movement	38	12	43	7
Separation movement	40	10	41	9
Movement of convergence continental-continental	42	8	46	4
Movement of convergence oceanic-continental	35	15	39	11
The earthquakes	41	9	45	5

expressed that students became more engaged during the post-test session because of how the information was presented. Students performed better when answering the practice exercises using ARGeo compared with the professor's lesson's answers. Moreover, combining mobile devices and AR causes students to lose track of time while learning about geography. AR causes enjoyment in students and a desire to repeat the experience as soon as possible. In summary, ARGeo is an alternative tool to learn about the natural components of the Earth, but it cannot replace the teacher.

7.4   CONCLUSIONS AND FURTHER WORK

Understanding what learning is and developing effective teaching–learning techniques are some of the most complex activities in the educational field. Learning means acquiring knowledge of something through study or experience. However, not all people learn in the same way. It is a fact that using technology in the classroom helps arouse students' interest and motivates them to acquire new knowledge.

The ARGeo prototype for helping fifth-grade students learn geography was presented in this chapter. ARGeo addressed the topic concerning the natural components of the Earth. The prototype allows fifth-grade students to interact and learn about the layers of the Earth, sliding, separation, continental–continental and oceanic–continental movements, and earthquakes. A total of 25 students participated in the study. The students completed a survey regarding motivation, quality, TAM, and ARI.

The results from the study revealed that students felt motivated to learn topics regarding the natural components of the Earth. Moreover, the students considered ARGeo a quality prototype and accepted its use in the educational setting. Students increased their grades when ARGeo was employed to learn and answer test questions. All this leads to the conclusion that ARGeo is a valuable complementary tool for learning the concepts related to the natural components of the Earth.

Extensions of the proposed study include adding audio to explain each topic to be learned. The use of audio will help students who prefer to learn auditorily.

Optimization of the display of virtual objects is required because, on some occasions, in mid-range mobile devices, the animations freeze. Also, running a pilot study using HMDs would be desirable to observe if the possibility of not clicking on screens increases students' motivation and achievement. Finally, it is planned to carry out a study to observe how AR is accepted in the long term. Currently, most AR studies benefit from the novelty of the technology. However, it would be interesting to know how long students' interest in using AR lasts.

REFERENCES

Acosta, M., González, S., Romero. M., Reza, L., Salinas. A., & Mendoza. K. (2019). *Geografía: Quinto Grado* (4th ed.). Dirección General de Materiales Educativos de la Secretaría de Educación Pública.

Adedokun, N., Ajani, A., Nuhu, K., & Shittu, A. (2020). Augmented Reality Instructional Tool in Enhancing Geography Learners Academic Performance and Retention in Osun State Nigeria. *Education and Information Technologies*, *25*(4), 3021–3033. https://doi.org/10.1007/s10639-020-10099-2

Barraza, R., Cruz, V., & Vergara, O. (2015). A Pilot Study on the Use of Mobile Augmented Reality for Interactive Experimentation in Quadratic Equations. *Mathematical Problems in Engineering*, *2015*, 1–13. https://doi.org/10.1155/2015/946034

Bernacki, M., Greene, J., & Crompton, H. (2020). Mobile Technology, Learning, and Achievement: Advances in Understanding and Measuring the Role of Mobile Technology in Education. *Contemporary Educational Psychology*, *60*, 1–8. https://doi.org/https://doi.org/10.1016/j.cedpsych.2019.101827

Brosvision s.r.o. (2023, February). *Augmented Reality Marker Generator – Brosvision*. www.brosvision.com/ar-marker-generator/

Bunge, W. (1973). The Geography. *The Professional Geographer*, *25*(4), 331–337. https://doi.org/10.1111/j.0033-0124.1973.00331.x

Chang, C., & Sheng, B. (2018). Teaching Geography with Technology—A Critical Commentary. In Chang, C. H., Wu, B., Seow, T., & Irvine, K. (Eds.), *Learning Geography Beyond the Traditional Classroom: Examples from Peninsular Southeast Asia* (pp. 35–47). Springer Singapore. https://doi.org/10.1007/978-981-10-8705-9_3

Coronado, A. (2020). *Aplicación Móvil de Realidad Aumentada para la Enseñanza de Geografía a Estudiantes de Primaria* [B.Sc. dissertation]. Universidad Autonoma de Ciudad Juarez.

Dalla, S., Di Nucci, D., Palomba, F., & Tamburri, D. (2020). Toward a Catalog of Software Quality Metrics for Infrastructure Code. *Journal of Systems and Software*, *170*, 1–8. https://doi.org/https://doi.org/10.1016/j.jss.2020.110726

Davis, F. (1989). Perceived Usefulness, Perceived Ease of Use, and User Acceptance of Information Technology. *MIS Quarterly*, *13*(3), 319–340. https://doi.org/10.2307/249008

Georgiou, Y., & Kyza, E. (2017). The Development and Validation of the ARI Questionnaire: An Instrument for Measuring Immersion in Location-Based Augmented Reality Settings. *International Journal of Human-Computer Studies*, *98*, 24–37. https://doi.org/https://doi.org/10.1016/j.ijhcs.2016.09.014

Hernández, L., López, J., Tovar, M., Vergara, O., & Cruz, V. (2021). Effects of Using Mobile Augmented Reality for Simple Interest Computation in a Financial Mathematics Course. *PeerJ Computer Science*, *7:e618*(1), 1–33. https://doi.org/10.7717/peerj-cs.618

Hon, C., & Feng, J. (2023). V-Model with Fuzzy Quality Function Deployments for Mobile Application Development. *Journal of Software: Evolution and Process*, *35*(1), 1–15. https://doi.org/10.1002/smr.2518

Jabangwe, R., Edison, H., & Duc, A. (2018). Software Engineering Process Models for Mobile App Development: A Systematic Literature Review. *Journal of Systems and Software*, *145*, 98–111. https://doi.org/https://doi.org/10.1016/j.jss.2018.08.028

Li, K., & Keller, J. (2018). Use of the ARCS Model in Education: A Literature Review. *Computers & Education*, *122*, 54–62. https://doi.org/https://doi.org/10.1016/j.compedu.2018.03.019

Liberatore, M., & Wagner, W. (2021). Virtual, Mixed, and Augmented Reality: A Systematic Review for Immersive Systems Research. *Virtual Reality*, *25*(3), 773–799. https://doi.org/10.1007/s10055-020-00492-0

Lin, P., Chai, C., Jong, M., Dai, Y., Guo, Y., & Qin, J. (2021). Modeling the Structural Relationship Among Primary Students' Motivation to Learn Artificial Intelligence. *Computers and Education: Artificial Intelligence*, *2*, 1–7. https://doi.org/10.1016/j.caeai.2020.100006

Loorbach, N., Peters, O., Karreman, J., & Steehouder, M. (2014). Validation of the Instructional Materials Motivation Survey (IMMS) in a Self-Directed Instructional Setting Aimed at Working with Technology. *British Journal of Educational Technology (BJET)*, *46*(1), 204–218. https://doi.org/10.1111/bjet.12138

Pan, J., Yang, X., Xie, X., Willis, P., & Zhang, J. (2009). Automatic Rigging for Animation Characters with 3D Silhouette. *Computer Animation & Virtual Worlds*, *20*(2–3), 121–131. https://doi.org/10.1002/cav.284

Piotrowska, I., Cichoń, M., Abramowicz, D., & Sypniewski, J. (2019). Challenges in Geography Education—A Review of Research Problems. *Quaestiones Geographicae*, *38*(1), 71–84. https://doi.org/doi:10.2478/quageo-2019-0009

Schultz, R., & DeMers, M. (2020). Transitioning from Emergency Remote Learning to Deep Online Learning Experiences in Geography Education. *Journal of Geography*, *119*(5), 142–146. https://doi.org/10.1080/00221341.2020.1813791

Sen, S., & Yildirim, I. (2022). A Tutorial on How to Conduct Meta-Analysis with IBM SPSS Statistics. *Psych*, *4*(4), 640–667. https://doi.org/10.3390/psych4040049

Shelton, B., & Hedley, N. (2002). Using Augmented Reality for Teaching Earth-Sun Relationships to Undergraduate Geography Students. *Proceedings of the First IEEE International Workshop Augmented Reality Toolkit*, *1*, 8. https://doi.org/10.1109/ART.2002.1106948

Soni, L., Kaur, A., & Sharma, A. (2023). A Review on Different Versions and Interfaces of Blender Software. *Proceedings of the 7th International Conference on Trends in Electronics and Informatics (ICOEI)*, 882–887. https://doi.org/10.1109/ICOEI56765.2023.10125672

Turan, Z., Meral, E., & Sahin, I. (2018). The Impact of Mobile Augmented Reality in Geography Education: Achievements, Cognitive Loads And Views Of University Students. *Journal of Geography in Higher Education*, *42*(3), 427–441. https://doi.org/10.1080/03098265.2018.1455174

Volioti, C., Keramopoulos, E., Sapounidis, T., Melisidis, K., Kazlaris, G., Rizikianos, G., & Kitras, C. (2022). Augmented Reality Applications for Learning Geography in Primary Education. *Applied System Innovation*, *5*(6), 1–25. https://doi.org/10.3390/asi5060111

Whitt, P. (2023). An Overview of GIMP 2.10. In Beginning Photo Retouching and Restoration Using GIMP: Learn to Retouch and Restore Your Photos like a Pro. Apress. https://doi.org/10.1007/978-1-4842-9265-5_1

8 Case Study II
Assemble an Educational Toy with Augmented Reality

8.1 INTRODUCTION

Product assembly is a crucial stage in manufacturing conducted after the production process. An assembly is the task of placing two or more pieces to create a product that meets the customer's specifications. However, there is a relationship between the complexity of the assembly and the number of errors generated when putting the pieces together. A new assembly is formed when one or more parts are assembled. Therefore, all assembly stages constitute the assembly sequence of a product (Cohen et al., 2019).

Product assembly is conducted by following step-by-step digital or printed instructions and diagrams explaining the structural relations between the product components. Instructions must include clear and intuitive assembly explanations. Moreover, many product instructions are poorly designed and out of date (Agrawala et al., 2003). Unfortunately, creating instructions that are easy to follow is challenging because mostly 2D flat views of the components are included.

The assembly task also must be done when buying a new product for the home. After opening the package, the instructions for assembly must be followed. However, the instructions are often unclear or extensive, or people deliberately ignore them. Following poorly designed instructions is cognitively demanding and causes frustration and confusion to many people (Heiser et al., 2004).

In the search to solve the problem, a short manual known as the "quick guide" is usually included. However, the quick guide is frequently too short and confusing. Therefore, for the assembly, there are three options: one is to read the manuals, the other is to look for videos in which assembly is explained, and the other is to ask for help from someone who knows how to assemble the product (Funk et al., 2018).

AR can help the product assembly process by providing step-by-step instructions supported by 3D or 2D virtual models (Rentzos et al., 2013). According to the paper by Bottani and Vignali (2019), most implementations of AR are related to assembly. The works by Agati et al. (2020) and Wang et al. (2022) can be consulted to learn about many fields that have employed AR for assembly purposes.

In this chapter, we present the results of the research conducted in the Ms.C. dissertation by González (2021), where an AR prototype was designed to present step-by-step instructions for assembling an educational toy. The System Usability Scale (SUS) (Brooke, 1996) and the time employed to conduct the assembly were measured to evaluate the usability and efficiency of the proposal.

As explained in Chapter 6, SUS states that the intention and validation of using technology can be measured by usability. Therefore, the SUS questionnaire comprising 10 questions with a scale of 1 to 5 was employed. The SUS scale is divided into the following intervals: 1–51, 51–68, 68–80.3, +80.3, and results higher than 68 are considered good quality (Vlachogianni & Tselios, 2022).

Accordingly, the following hypothesis was posed:

- H_1: Using AR in assemblage tasks can achieve a usability value higher than 68.

On the other hand, the time taken to conduct the assembly is also another critical factor. Making an assemblage without any informational support can impact time. Therefore, the time to conduct the assembly with and without AR was recorded. A statistical t-test was employed to determine whether significant differences existed between the recorded times (Tae, 2015). In addition, although a study was not conducted in this regard, the mental load and assembly errors are expected to decrease using the prototype.

As a result, the following hypothesis was posed:

- H_2: Using AR in assemblage tasks will reduce the completion time compared to traditional means.

The rest of this chapter is organized as follows. In Section 8.2, we present the methodology to build the prototype. The experiments and results are shown in Section 8.3. Finally, the conclusions are presented in Section 8.4.

8.2 DESIGN OF THE AUGMENTED REALITY PROTOTYPE FOR DRAGSTER ASSEMBLY

The AR prototype was designed under the following premise: "If we use a set of markers and three-dimensional models, then we can design a system that induces the user to assemble an educational toy correctly." The educational toy selected was a dragster because it includes the basic principles of physics and electronics.

The methodology to design the AR prototype shown in Figure 8.1 comprises five main stages: (i) camera tracking system, (ii) video mixing system, (iii) interaction, (iv) usability testing, and (v) time testing.

8.2.1 CAMERA TRACKING SYSTEM

The tracking system scans and models 3D objects. In addition, feature extraction, feature matching, and homography calculation are conducted for tracking. A dragster was selected as the educational toy to conduct the assembly. Therefore, the first task involves identifying all the parts to conduct the assembly. Moreover, it was decided that each part would serve as a marker. Table 8.1 lists all the items needed to assemble the toy, and Figure 8.2 depicts all the items.

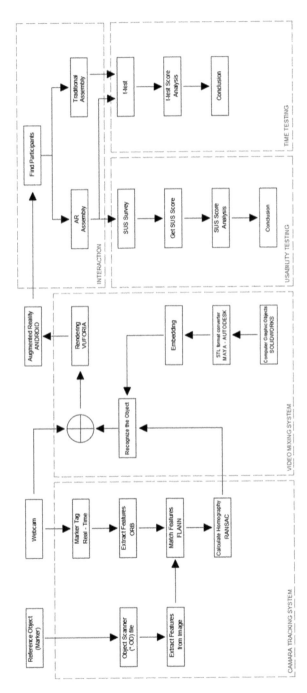

FIGURE 8.1 The scheme of the methodology to design the AR prototype for assembly.

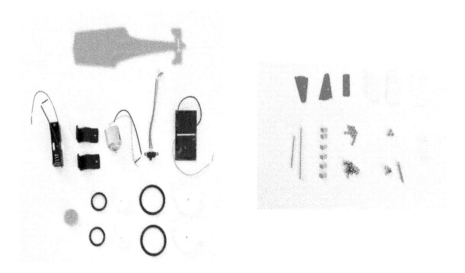

FIGURE 8.2   The items needed to assemble the dragster.

Each toy item was scanned and stored in a database using the Vuforia Object Scanner (VOS). As a result, a *.OD file was generated. Scanning collects data regarding an object's appearance (Garner et al., 2011). The most important part of the scan is to visualize the number of points extracted. The more components the object has, the more points are generated.

Consequently, the model will be displayed with more quality. Each 3D model will be superimposed in the AR scene. In addition to scanning, images of each piece were acquired to be used as markers.

The features of each image were extracted with the Oriented FAST and Rotated BRIEF (ORB) algorithm (Rublee et al., 2011). ORB was selected because it is free to use, unlike SIFT and SURF. ORB uses the Features from Accelerated Segment Test (FAST) method for the first stage of feature extraction (Viswanathan, 2009). FAST is employed to discard useless points and strengthen those identified as "corners." The Binary Robust Independent Elementary Features (BRIEF) generates the features in the second stage (Michael et al., 2010). The features extracted were stored in a database.

After feature extraction, the matching process starts. The matching is conducted using the FLANN algorithm. Finally, the homography is calculated using RANSAC (consult Chapter 3). All the processes of the camera tracking system were conducted offline.

8.2.2 Video Mixing System

Once all parts are characterized, a starting point to conduct the assembly must be determined. Therefore, the car's yellow base was considered the "mother piece" because all the items will be mounted on it.

The video mixing system recognizes the object using extracted features and draws the AR scene. In addition, SolidWorks was employed to design and test the virtual assembly of the dragster (Lombard, 2013). SolidWorks allows better visualization

TABLE 8.1
The List of All Items Needed to Assemble the Dragster

Item No.	Description	Quantity
P1	Engine	1
P2	Gear for engine	1
P3	Face gear	1
P4	Frontal metal shaft	1
P5	Rear metal shaft	1
P6	Short screws	12
P7	Large screw	1
P8	M2 Nuts	14
P9	M3 Nuts	3
P10	8 mm Screw	2
P11	20 mm Screw	1
P12	Metal plates	6
P13	Frontal wheels	2
P14	Rear wheels	2
P15	Frontal rubber tire	2
P16	Rear rubber tire	2
P17	Small PVC tubes	4
P18	Big PVC tubes	2
P19	Motor holder	2
P20	Panel holder	1
P21	Large sticky sponge	1
P22	Small sticky sponge	1
P23	Solar panel	1
P24	Wire tie	1
P25	Switch	1
P26	Battery compartment	1
A	Automobile base	1
B	Plastic mold	1
C	Fins	2

and free part manipulation, determining the assembly's geometric restrictions. Then, Maya software was employed for fine-tuning design details. The screws nuts, and gear specifications were generated with Maya (Derakhshani, 2012). It is essential to highlight that in this step, the *.STL extension was preserved. After an exhaustive analysis, four main assembly steps were determined to assemble the dragster fully.

In the first assembly, the user must locate the mother piece. Afterward, the red plastic mold corresponding to the front wheels must be screwed to the mother piece with an 8 mm screw and a nut (M2). Then, the motor holders should be installed on the mother piece using two short screws (8 mm) and two nuts (M3). Finally, two metal plates must be added to the back of the mother piece, securing them with two short screws (6 mm) and two nuts (M2). Figure 8.3 shows the real and virtual views of the first assembly.

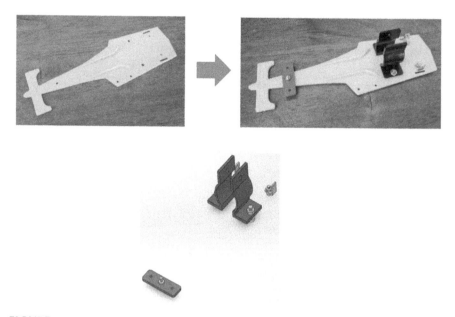

FIGURE 8.3   The first assembly step.

For the second assembly, the user must locate the red plastic mold corresponding to the front wheels and install two metal plates at the ends. The plates are fixed using two short screws (6 mm) and two nuts (M2). Then, the motor gear must be placed on the mother piece and between the holders. Finally, the red fins are attached at the back of the mother piece, securing them using two metal plates, two short screws (6 mm), and two nuts (M2). Figure 8.4 shows the real and virtual views of the second assembly.

The frontal metal shaft must be inserted in the red plastic mold corresponding to the front wheels for the third assembly. Place the frontal wheels and the frontal rubber tires around the wheels. Then, insert the rear metal shaft in the center machinery through the red plastic tabs, ensuring both gears align. Cover the shaft with a small PVC tube on each end. Place the rear wheels as well as the rubber tires around the wheels. Place two metal plates on top of the fins. Secure them using two short screws (6 mm) and two nuts (M2). Finally, place a small sticky sponge in the middle of the mother piece. Figure 8.5 shows the real and virtual views of the third assembly.

Secure the motor using a large screw (25 mm) and a nut (M3) for the fourth assembly. On top of the red fins, locate the panel clip and secure it using two short screws (6 mm). Place a large sticky sponge on top of the panel holder and place the solar panel on top of it. Then, place the battery compartment in the central part of the mother piece. Finally, install the switch using two short screws (6 mm) and two nuts (M2). Figure 8.6 shows the real and virtual views of the fourth assembly. Moreover, different views of the assembled dragster are shown in Figure 8.7.

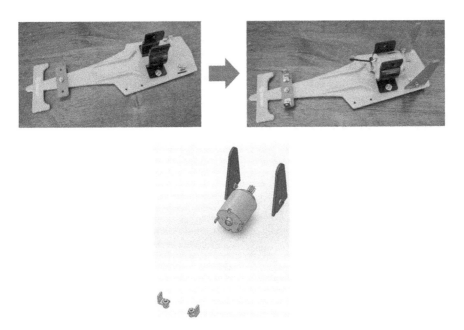

FIGURE 8.4 The second assembly step.

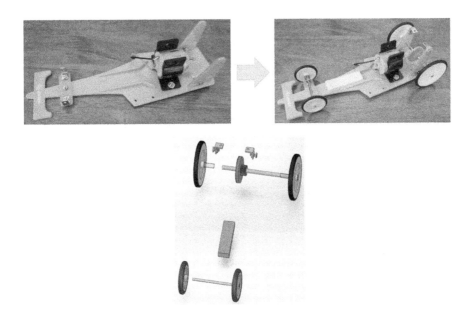

FIGURE 8.5 The third assembly step.

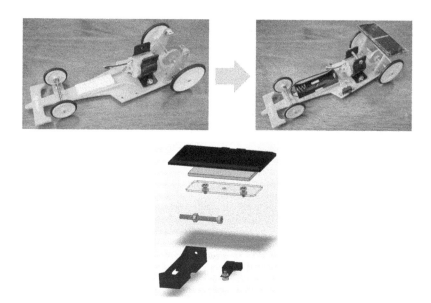

FIGURE 8.6   The fourth assembly step.

8.2.3 INTERACTION

FIGURE 8.7   Different views of the dragster.

The interaction stage involves finding the participants to conduct the dragster assembly with traditional and AR instructions. All the participants were informed about the goal of the study. Then, the dragster assembly using printed instructions was started. Users were provided with all the necessary tools to conduct the test. After finishing the assembly with the traditional method, the users started the assembly with the AR prototype.

The AR prototype was built using Unity and Vuforia Software Developer Kit (SDK). Moreover, all the computer vision techniques were programmed using the OpenCV library. The prototype runs on mobile devices with an Android

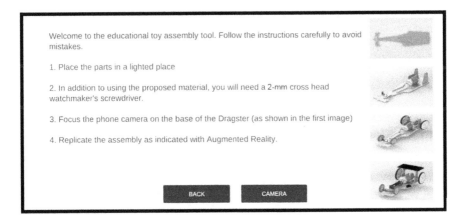

FIGURE 8.8   The instructions to interact with the AR prototype.

operating system. We provided a mobile device to each user aiming to experiment with AR.

The interaction with AR begins when the user holds the tablet and taps the dragster icon to start the execution. The presentation screen is displayed, followed by a screen with basic instructions to interact with the prototype. Figure 8.8 shows the screen with the instructions.

After reading the instructions, the camera button must be pressed. Immediately, the mobile device's camera is turned on. The user must point to the mother piece and follow the instructions to complete the dragster assembly. The time employed for each assembly was recorded, and finally, each participant was asked to fill out the SUS questionnaire.

8.2.4   USABILITY TESTING

The SUS survey containing 10 questions was employed to acquire users' comments regarding the usability of the AR prototype. The survey was printed and delivered to each participant. An average time of 15 minutes was required to complete the survey. The responses from each user were captured and stored in a database. A double-check was performed to ensure that the captured data were correct. Fortunately, no outliers were detected.

Once the data were captured, an analysis of the answers offered by each user was conducted. This stage was carried out following the explanations offered in Chapter 6. Finally, the SUS score was computed, and a conclusion regarding the findings was obtained.

8.2.5   TIME TESTING

The time employed by each participant to conduct each assembly was recorded in a database. Two people helped to take the time to avoid errors in the measurement. All the measurements were captured and stored in a database. A double-check was performed to ensure that the captured data were correct. Fortunately, no outliers were detected.

Afterward, a *t*-test was conducted to determine whether a significant difference exists between the time recorded in traditional assembly and the time recorded for assembly using the AR prototype. This stage was carried out following the explanations in Chapter 6. Finally, a conclusion regarding the findings was obtained.

8.3   EXPERIMENTS AND RESULTS

All the experiments were conducted in August 2021. Because of the pandemic, the student's house in CDMX, Mexico, was employed as the setting to conduct the experiments. Twenty persons (10 male and 10 female) from ages 8 to 77 (M = 38.95, SD = 20.58) tested the AR prototype. Three sessions were organized with seven, seven, and six participants. The sessions were conducted two days apart and lasted 3 hours. The participants were primary, secondary, high school, undergraduate students, homemakers, retired adults, and parents.

A quasi-experimental study established a cause-and-effect relationship between independent and dependent variables. Therefore, the sample to study was not randomly selected, and no control group was required (Furtak et al., 2012). Participants were informed about the research goal and that the data obtained would be treated with confidentiality and used only for academic purposes.

All participants conducted the assembly dragster task using the traditional method and the AR prototype. The time employed for each assembly was recorded, and finally, each participant was asked to fill out the SUS survey. The survey was printed and delivered to each participant. An average time of 15 minutes was required to

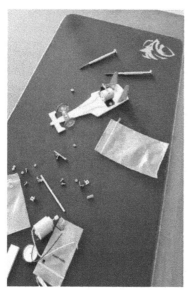

FIGURE 8.9   The setting to conduct the experiments.

complete the survey. Figure 8.9 shows an example of the setting employed to conduct the experiments.

8.3.1 USABILITY TESTING RESULTS

The SUS survey includes 10 questions that must be rated on a Likert scale from 1 to 5 (1: strongly disagree, 2: disagree, 3: neutral, 4: agree, and 5: strongly agree).

The 10 questions of the SUS survey can be consulted in Chapter 6. Each of the items receives a score (1–5). The algorithm to compute the SUS score is the following:

1. Sum up the total for the odd items (1, 3, 5, 7, 9).
2. Subtract five from the total to get the odd score.
3. Sum up the total for the even items (2, 4, 6, 8, 10).
4. Subtract the total from 25 to get the even score.
5. Sum up the even and odd scores.
6. Multiply the sum by 2.5 to obtain the final SUS score.

Table 8.2 shows the data obtained from each participant after testing the AR proto-type. Data collected were captured to generate a database with IBM SPSS software (Sen & Yildirim, 2022). After collecting and capturing the data, a normality test with a

TABLE 8.2
Data Collected from the SUS Survey and Time for Assembly

ID	Sex	Age	AR time (m)	Traditional time (m)	SUS Questionnaire									
					1	2	3	4	5	6	7	8	9	10
P1	F	57	38	47	4	4	3	4	4	2	3	1	3	1
P2	M	60	24	31	4	1	4	2	5	3	4	2	4	2
P3	F	30	33	39	4	4	5	3	4	2	4	2	4	2
P4	M	8	39	42	5	2	4	2	5	2	3	3	5	3
P5	F	24	41	36	5	1	4	2	4	4	4	2	4	2
P6	M	45	28	31	4	2	3	2	5	3	5	2	5	1
P7	F	28	46	51	4	2	3	4	4	1	3	5	3	2
P8	M	33	22	25	5	3	4	1	5	1	5	3	4	1
P9	F	15	29	24	5	2	3	3	4	2	3	1	4	2
P10	M	35	34	32	4	4	4	3	5	4	4	3	3	2
P11	F	9	58	61	5	1	5	3	4	3	5	3	3	2
P12	M	77	48	39	4	2	4	2	4	2	4	2	4	2
P13	F	72	65	74	2	3	3	3	3	2	3	1	3	4
P14	M	58	47	52	4	3	3	2	4	3	4	3	4	2
P15	F	51	36	30	4	2	4	2	5	4	2	1	4	1
P16	M	19	44	41	4	2	5	2	4	3	5	3	5	1
P17	F	24	29	33	5	2	4	4	5	4	5	3	4	2
P18	M	28	20	19	2	4	3	3	2	4	4	4	2	3
P19	F	44	32	36	5	2	4	1	5	4	4	1	5	1
P20	M	62	53	49	4	2	2	3	4	4	2	2	4	2

confidence level of 95% was performed. Since the sample is less than 50, the Shapiro-Wilk test was conducted. The normality test indicated that data from the survey were not normally distributed. The result is correct because, for a good score in the SUS, it is desired to obtain a minimum score in the even-numbered item. Meanwhile, in the odd-numbered items, a score close to 5 is expected.

The equations to compute the SUS score were programmed using Python 3.8. All the statistics computed are shown in Table 8.3. As can be observed, the SUS score obtained was 69.37. Although the SUS scores range from 0 to 100, they should not be considered a percentage. Therefore, a way to interpret the results is to normalize the scores to produce a percentile ranking.

Lewis and Sauro (2018) analyzed data from over 5,000 users across 500 different evaluations and concluded that SUS is a reliable and valid measure of perceived usability. Moreover, from the analysis, a curved grading scale was generated in which a SUS score of 68 is at the center of the range. A result above 68 is considered good. A result below 68 indicates that several aspects of the system must be corrected.

Percentile ranks inform how well the raw scores compare to others in the database. Therefore, the average score at the 50th percentile is 68. Bangor et al. (2009) conducted a study that included one more SUS item to answer the following question: What is the absolute usability associated with any individual SUS score?

A seven-point Likert scale was used to determine a phrase associated with a range of SUS scores. Therefore, the rating scale includes the following adjectives: worst imaginable (12.5), awful (20.3), poor (35.7), okay (50.9), good (71.4), excellent (85.5), and best imaginable (90.1).

The worst evaluation was for Participant #18 (28 years old), with 37.5 in the SUS evaluation, resulting in "poor." On the other hand, the best evaluation was for Participant #8 (33 years old) and Participant #19 (44 years old), both with a result of 85 in the SUS evaluation, resulting in "excellent."

Figure 8.10 shows the index of adjectives obtained; "awful" and "okay" did not get results, while "excellent" is in third place with two results. "Poor" is in the second position with seven results, and "good" is the most frequent with 12 results.

Once the results were obtained, a t-test was conducted to observe if the difference between 68 and 69.37 in the SUS score was significant. The study started by defining the means, the variance, and the number of samples: $\bar{x}_1 = 68$, $\bar{x}_2 = 69.375$, $s_c^2 = 68.708$, and $n_1 = n_2 = 20$. The statistical test value was computed using equation 8.1.

TABLE 8.3
Statistics Computed from the SUS Survey

Statistic	Value
Standard deviation	11.72
Median	72.5
Lowest evaluation	37.5
Highest evaluation	85.0
Mean	69.37

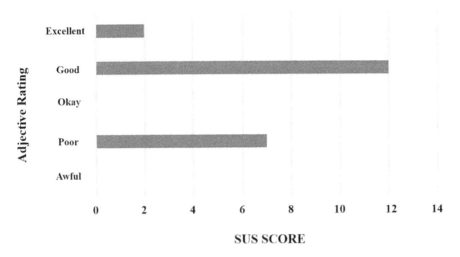

FIGURE 8.10   The adjective rating for SUS.

$$t = \frac{68 - 69.375}{\sqrt{\dfrac{68.708}{20} + \dfrac{68.708}{20}}} = -0.52 \qquad (8.1)$$

Then, degrees of freedom, $DoF = n_1 + n_2 - 2 = (20 + 20 - 2) = 38$, are computed. Therefore, the critical value with a significance level of 5% is:

$$t_{\left(1-\frac{\infty}{2}\right)(n_1 + n_2 - 2)} = 2.0243 \qquad (8.2)$$

On the t-distribution curve, the critical value is 2.0243 to the left and right, while the calculated value is in the opposite region of acceptance for H_0. Therefore, H_1 is accepted: "Using AR in assemblage tasks can achieve a usability value higher than 68." The best score obtained was 85, while the lowest evaluation was 37.5. The average obtained was 69.375, which is greater than 68, which indicates the acceptance of the usability of the designed prototype.

8.3.2   TIME TESTING RESULTS

Twenty participants executed the assemblage using an AR prototype and the traditional method. Therefore, the study was conducted to discover whether AR improves performance time. Table 8.2 shows each participant's time to conduct the experiment.

The test of Shapiro-Wilk was employed to select the statistical analysis tool according to the data distribution. The results obtained with a 5% significance level using SPSS indicated that data sets follow a normal distribution. Table 8.4 shows the statistical data obtained regarding time for both assembles.

TABLE 8.4
Statistics Regarding Time to Experiment

Statistic	AR time	Traditional time
Mean	38.3	39.6
Median	37	37.5
Variance	145.37	174.98
Standard deviation	12.05	13.22
Minimum	20	19
Maximum	65	74
Range	45	55
Skewness	0.518	0.923
Kurtosis	-0.196	1.111

Finally, a paired t-test with a 5% significance level was calculated. First, the degrees of freedom, $DoF = (n-1) = (20-1) = 19$, are computed. Then, the critical value is computed using equation 8.3.

$$t_{(1-\infty)(n-1)} = 1.729 \tag{8.3}$$

The critical value must be compared to the statistical test value using equation 8.4.

$$t_s = \frac{\bar{d}}{\frac{s_d}{\sqrt{n}}} = \frac{39.6 - 38.3}{\frac{1.1710}{\sqrt{20}}} = 4.9646 \tag{8.4}$$

Finally, the p-value $(0.000043) < \alpha = 0.05$ was obtained. The critical value (**1.729**) is the limit of the acceptance region. Therefore, the statistical value (**4.9646**) is within the rejection region of the null hypothesis (H_0). Therefore, the null hypothesis should not be accepted.

The difference of 1.3 between means is statistically significant ($p < 0.001$). Therefore, H_2 is accepted, which means using AR takes less time to assemble the dragster than the traditional method.

8.4 CONCLUSIONS

This chapter presented the design and test of an AR prototype for helping assemble a dragster. Two hypotheses were posed, the first regarding prototype usability and the second regarding the time employed to conduct the assembly. The results revealed that the AR prototype is an interactive tool with good usability and an improvement in completion time. The prototype is also relevant in assembly tasks, promoting comfort and fun. Moreover, it is unnecessary to be an expert or to fully understand a topic for making the assembly.

Although the project has been concluded, improvements could continuously be developed. AR for assemblage can be improved by creating larger-scale models. Currently, the application is being developed in English. Its translation into different languages can extend usability for more people. According to the results obtained, it was found that people over 45 rejected using our technology. Then, as part of this testing process, a stage was implemented to improve or redesign the assembly tool. That was due to feasibility and can be attributed to the objects' manipulation. The feasibility problem is the manipulation between holding the mobile device with one hand and carrying out the assemblage with the other. Therefore, it is also essential to carry out another study where we leave the user's hands free using AR glasses.

REFERENCES

Agati, S., Bauer, R., Hounsell, M., & Paterno, A. (2020). Augmented Reality for Manual Assembly in Industry 4.0: Gathering Guidelines. *Proceedings of the 22nd Symposium on Virtual and Augmented Reality (SVR)*, 179–188. https://doi.org/10.1109/SVR51698.2020.00039

Agrawala, M., Phan, D., Heiser, J., Haymaker, J., Klingner, J., Hanrahan, P., & Tversky, B. (2003). Designing Effective Step-by-Step Assembly Instructions. *ACM Transactions on Graphics*, *22*(3), 828–837. https://doi.org/10.1145/882262.882352

Bangor, A., Kortum, P., & Miller, J. (2009). Determining What Individual SUS Scores Mean: Adding an Adjective Rating Scale. *Journal of Usability Studies*, *4*(3), 114–123.

Bottani, E., & Vignali, G. (2019). Augmented Reality Technology in The Manufacturing Industry: A Review of The Last Decade. *IISE Transactions*, *51*(3), 284–310. https://doi.org/10.1080/24725854.2018.1493244

Brooke, J. (1996). Sus: A Quick and Dirty Usability. *Usability Evaluation in Industry*, *189*(3), 189–194.

Cohen, Y., Faccio, M., Pilati, F., & Yao, X. (2019). Design and Management of Digital Manufacturing and Assembly Systems in The Industry 4.0 Era. *International Journal of Advanced Manufacturing Technology*, *105*(9), 3565–3577. https://doi.org/10.1007/s00170-019-04595-0

Derakhshani, D. (2012). *Introducing Autodesk Maya 2013*. John Wiley & Sons.

Funk, M., Lischke, L., Mayer, S., Sahami, A., & Schmidt, A. (2018). Teach Me How! Interactive Assembly Instructions Using Demonstration and In-Situ Projection. In Huber, J., Shilkrot, R., Maes, P., & Nanayakkara, S. (Eds.), *Assistive Augmentation* (pp. 49–73). Springer Singapore. https://doi.org/10.1007/978-981-10-6404-3_4

Furtak, E., Seidel, T., Iverson, H., & Briggs, D. (2012). Experimental and Quasi-Experimental Studies of Inquiry-Based Science Teaching: A Meta-Analysis. *Review of Educational Research*, *82*(3), 300–329. https://doi.org/10.3102/0034654312457206

Garner, R., Blackburn, S., & Frampton, D. (2011). A Comprehensive Evaluation of Object Scanning Techniques. *Proceedings of the International Symposium on Memory Management*, 33–42. https://doi.org/10.1145/1993478.1993484

González, D. (2021). *Augmented Reality Based on Assembly Instructions for an Educational Toy* [M.. Sc. dissertation]. Centro de Investigación en Computación (CIC) del Instituto Politécnico Nacional (IPN).

Heiser, J., Phan, D., Agrawala, M., Tversky, B., & Hanrahan, P. (2004). Identification and Validation of Cognitive Design Principles for Automated Generation of Assembly

Instructions. Proceedings of the Working Conference on Advanced Visual Interfaces, 311–319. https://doi.org/10.1145/989863.989917

Lewis, J., & Sauro, J. (2018). Item Benchmarks for the System Usability Scale. *Journal of Usability Studies, 13*(3), 158–167.

Lombard, M. (2013). *SolidWorks 2013 Bible*. John Wiley & Sons.

Michael, C., Lepetit, V., Strecha, C., & Fua, P. (2010). BRIEF: Binary Robust Independent Elementary Features. In Daniilidis, K., Maragos, P., & Paragios, N. (Eds.), *Proceedings of the European Conference on Computer Vision (ECCV)* (pp. 778–792). Springer Berlin Heidelberg.

Rentzos, L., Papanastasiou, S., Papakostas, N., & Chryssolouris, G. (2013). Augmented Reality for Human-based Assembly: Using Product and Process Semantics. *IFAC Proceedings* Volumes, *46*(15), 98–101. https://doi.org/https://doi.org/10.3182/20130 811-5-US-2037.00053

Rublee, E., Rabaud, V., Konolige, K., & Bradski, G. (2011). ORB: An efficient alternative to SIFT or SURF. *Proceedings of the International Conference on Computer Vision (ICCV)*, 2564–2571. https://doi.org/10.1109/ICCV.2011.6126544

Sen, S., & Yildirim, I. (2022). A Tutorial on How to Conduct Meta-Analysis with IBM SPSS Statistics. *Psych, 4*(4), 640–667. https://doi.org/10.3390/psych4040049

Tae, K. (2015). T test as a Parametric Statistic. *Korean Journal of Anesthesiology, 68*(6), 540–546. https://doi.org/10.4097/kjae.2015.68.6.540

Viswanathan, D. (2009). Features from Accelerated Segment Test (FAST). *Proceedings of the 10th Workshop on Image Analysis for Multimedia Interactive Services*, 6–8.

Vlachogianni, P., & Tselios, N. (2022). Perceived Usability Evaluation of Educational Technology Using the System Usability Scale (SUS): A Systematic Review. *Journal of Research on Technology in Education, 54*(3), 392–409. https://doi.org/10.1080/15391 523.2020.1867938

Wang, Z., Bai, X., Zhang, S., Billinghurst, M., He, W., Wang, P., Lan, W., Min, H., & Chen, Y. (2022). A Comprehensive Review of Augmented Reality-Based Instruction in Manual Assembly, Training and Repair. *Robotics and Computer-Integrated Manufacturing, 78*, 102407. https://doi.org/https://doi.org/10.1016/j.rcim.2022.102407

9 Augmented Reality Challenges and Trends

9.1 INTRODUCTION

Over the last three and a half years (2020–mid-2023), the world has experienced changes like never before. Due to the pandemic, the medical field had to accelerate its traditional procedures for creating COVID-19 vaccines. In the technological field, people have changed their ways of working and personal interactions by using video-conferencing systems daily. In short, the pandemic forced the world to become more technological, and what is a fact is that things will not go back to the way they were before.

The AR field has also benefited from the growing use of technology. Today, new opportunities for using AR in various application domains have emerged. Moreover, AR will be consolidated as a daily use technology in a few years.

In the following sections, the AR challenges and trends are discussed. A challenge is considered as something that is difficult and needs great mental effort to be done successfully. In comparison, a trend is considered a direction of change in how something behaves.

9.2 AR CHALLENGES

Even though AR has developed considerably, many challenges must be resolved for its consolidation. In the state-of-the-art, various novel works have described the principal challenges that AR must face. The works by Danielsson et al. (2020), Rejeb et al. (2021), and Zhan et al. (2020) discuss the AR challenges regarding technological displays. The investigations of Alalwan et al. (2020), Alzahrani (2020), and Doerner and Horst (2022) describe the challenges in educational settings. In contrast, the research conducted by Dirin and Laine (2018), Masood and Egger (2019), and Qiao et al. (2019) have explained the general challenges in applying AR. Therefore, this section explains and groups technological, usability, and social challenges.

9.2.1 Technological Challenges

An AR system has to deal with a vast amount of information. Therefore, most AR experiences' success lies in the delivery device's features. The hardware devices should

DOI: 10.1201/9781003435198-9

be small, light, easily portable, and fast enough to display graphics. Unfortunately, most of the devices on the market that offer good features for implementing RA are usually very expensive. So, one of the leading technological challenges is to make the hardware to implement AR cheaper (Billinghurst, 2021). Therefore, creating a compact, lightweight, see-through display with sufficient brightness and contrast, a variable focal plane, a wide field of view, robust tracking, and high resolution is desirable (Zhan et al., 2020).

The real world is experienced with all the human senses, and similar to real objects, virtual objects in AR have non-visual and multimodal qualities. Therefore, it will be desirable that AR engages all the senses to create multimodal experiences that can lead, for example, to experience with AR with closed eyes. Multimodal AR facilitates the interaction of users with AR content (Kim et al., 2021).

Currently, AR users must learn how to manipulate virtual content using a set of cues such as markers. Therefore, it is recommended to design tangible AR interfaces that allow users to interact with virtual content using the same techniques as they would with a real physical object (Billinghurst et al., 2008). Prototyping tangible AR applications is challenging because of the need to register the physical properties in the AR scene, use the properties to provide haptic feedback, and enable the manipulation of virtual objects using the physical properties (Villanueva et al., 2022).

Researchers are currently working on designing intuitive interactions with AR applications by recognizing hand gestures. However, the set of gestures that can be recognized is still limited. Using gestures can reduce the user's mental load and is desirable in noisy environments where voice commands cannot be employed (Billinghurst et al., 2014). The goal is to design systems that can recognize various dynamic and static hand gestures, and consequently, the time to complete a task can be reduced.

Another field that would be important to explore is the combination of AR with brain–computer interfaces (BCIs). A BCI can provide an additional communication channel and a new way to control and interact with AR content. BCIs and AR can be used to design immersive scenarios through induced illusions of an artificially perceived reality (Angrisani et al., 2020). AR glasses can provide visual stimuli, and a single-channel electroencephalography (EEG) device can measure the evoked potentials for a suitable combination. In this regard, many technological challenges must be faced, such as information fusion, stimulus design, hardware synchronization, and dealing with distractions (Chen et al., 2020).

Involving many users in the same AR reality experience would be desirable. The goal is for AR users to stop being in extreme isolation and become part of a community known as collaborative AR. AR must deal with pose estimation and point-to-point networks to make local or remote collaborative systems (Marques et al., 2022). In a collaborative AR environment, the user can move freely with a controlling and independent point of view. Since the objects are shared, the same model must be registered and observed by all the users. The user–user and user–object interactions must be handled by designing robust protocols. Also, the object occlusion must be managed, and robust camera calibration techniques must be designed (García et al., 2020).

Advanced computer vision methods can enhance the functionality and effectiveness of AR applications. Unfortunately, most mobile devices on the market lack the hardware features to implement the most novel and potent vision algorithms. Therefore, creating ad hoc models to run on mobile devices will be challenging. Also, devices' processing power and storage capabilities are expected to increase significantly in a few years. Deep learning technology can reinforce AR apps by simulating the human learning method by learning through examples. Lightweight convolutional neural network (CNN) architectures can be implemented in mobile devices for solving tasks such as object segmentation, classification, and recognition (Lampropoulos et al., 2020).

Since AR employs cameras and sensors to project virtual content, the designers must keep the device's battery life in check. Moreover, overheating must be solved to run AR efficiently on all devices. Reducing the frame rate, decreasing the number of virtual objects to be processed and displayed, and creating half-standard resolution content can help to overcome the challenges of battery operating time and overheating in mobile devices (Ateya et al., 2023; Zhang et al., 2020).

9.2.2 USABILITY AND ERGONOMICS CHALLENGES

Despite the great potential of AR technology to be applied in various contexts, usability and ergonomics challenges must be surpassed to ensure daily life inclusion. Developing AR applications requires different design and development considerations than traditional desktop solutions. The lack of guidelines for the successful design and usability of AR applications is one of the main reasons that limits its use. No standard has been established yet regarding designing user interfaces and interactions for AR applications. Moreover, at the time of writing this book, the creation of standards for AR was still a little exploited topic. Without sufficient criteria, no developing company can secure the compatibility of AR with the wide variety of mobile devices available on the market (Datcu et al., 2015).

Minaee et al. (2022) mentioned that the models developed to create AR applications work well only under certain specifications. Therefore, new models that run in changing and complex environments are required. On the other hand, most AR models have been developed in two-dimensional images. However, to provide a more realistic experience, working with 3D models and more detailed textures is necessary.

Currently, it is not easy to create and distribute AR content. However, according to Muñoz et al. (2020), the popularity and usability of AR will increase due to the evolution of cloud technologies. This combination is named AR cloud and implies the digitization of the entire world. AR cloud will provide data and services directly related to the user's physical surroundings. Therefore, the real world is expected to become a shared screen space to promote the participation and collaboration of multiple users. The AR cloud is expected to be a digital 3D copy of the real world (Huang et al., 2012).

The lack of longitudinal studies is a challenge that must be faced to promote the everyday use of AR. Many measurements must be made throughout the process to identify the benefits effectively. Unfortunately, most AR apps are directed at entertainment and education, and their effectiveness has not been fully proven because only information on the first experience is recorded (Garzón, 2021). Many users in educational settings have expressed that the use of AR apps is complex. Therefore, it seems that AR applications are mainly used for their novelty and not for their true benefit in the field of application. It is advisable to carry out long-term studies, for example, during a school year, to determine when the novelty stops being important, and users use it because they find a true benefit. Attentional tunneling describes the problem that users focus more on the virtual content than on the topic that needs to be experienced (Syiem et al., 2021)

On the other hand, AR introduces ergonomic issues that must be confronted, such as musculoskeletal problems, visual clutter, occlusion, distraction, and privacy. For example, using an AR HMD over a workday causes visual fatigue and impacts concentration performance. Moreover, using an HMD in the industry can cause accidents because of the limited field of view. Also, using display information devices significantly increases the cognitive workload of AR users and can induce motion sickness (Kaufeld et al., 2022). Therefore, a challenge is minimizing the discomfort involved in AR displays.

According to Brown et al. (2023), ergonomic and human factors such as mental workload, situational awareness, information management, practical training, and general usability must be considered when an AR application is designed. Moreover, AR apps must always offer the best visualization of the field addressed. A challenge will be creating surveys to measure AR apps' ergonomics, such as the Augmented Reality Sickness Questionnaire (ARSQ) (Hussain et al., 2023).

The user's interaction with the AR environment is another ergonomic challenge to consider when designing applications. Performing repetitive movements with hands and fingers can cause fatigue and stress in the neck and shoulders, and those musculoskeletal problems could become chronic if the user is exposed to these activities for an extended period (Kim et al., 2020).

9.2.3 SOCIAL CHALLENGES

Currently, AR still does not receive full public acceptance. People still have no idea how to use AR technology daily because it is still considered a technology that has not fully matured or been sufficiently tested. However, as AR technologies develop and become more accessible, their impact on society will become evident.

One of the challenges to the social acceptance of AR is security privacy. The AR prototype must preserve the data's confidentiality, integrity, and availability and defend against malicious applications. Therefore, AR prototypes must clearly explain how the information is used, where the data are stored, and whether the data will be shared with third-party customers (Alismail et al., 2023).

AR employs data from various sensors to which unauthorized persons may have access. For example, a user's exact location can be shared in real time, and the information can be employed to cause harm to that person. On the other hand, a malicious application can trick AR users into causing a conflict. For example, overlay an incorrect speed limit on top of a real speed limit sign or locate a fake sign where it should not appear. Also, malicious applications can cause user sensory overload by flashing bright lights on the device screen, playing loud sounds, or delivering intense haptic feedback such as device vibrations (Roesner et al., 2014).

Developing successful and high-quality AR prototypes requires advanced 3D modeling, content design, programming, and computer vision skills. Unfortunately, industries, companies, and educational institutions often do not have appropriately qualified employees to develop AR prototypes. Moreover, decision-makers in companies are frequently not informed of AR's benefits. Therefore, a challenge is promoting the use of AR and creating platforms that facilitate the development of AR prototypes for almost anyone. Several AR toolkits have been introduced, such as ARkit by Apple (Wang, 2018) and ARCore by Google (Lanham, 2018), creating a more accessible and friendlier approach to designing AR apps.

The ethical implications of using AR are another social challenge that must be addressed. The first thing that must be guaranteed is the responsible use of AR applications. Excessive use of AR could lead to decreased real-world interactions and a greater sense of isolation. AR should be used not to overwhelm users, and boundaries should be defined to ensure it is used appropriately (Neely, 2019). Users should be informed that excessive use of AR devices can impact physical health by straining eyes and posture.

Another challenge is finding strategies to prevent companies from using AR to make their products appear more attractive than they are. False information could mislead consumers into purchasing decisions they would not otherwise have made. AR should not be used to create immersive experiences that are too intense, or that could potentially cause distress and mental health effects. Therefore, companies developing AR technology must be transparent about the potential risks and benefits associated with its use (Sabelman & Lam, 2015).

9.3   AR TRENDS

In the literature, various authors have contributed to investigating and monitoring the development of AR over the years to anticipate trends in various areas such as education, entertainment, health, and industry (Altinpulluk, 2019; Bacca et al., 2014; Devagiri et al., 2022; K. Kim et al., 2018; Muñoz et al., 2020).

This section discusses the trends regarding the industry, education, multimodal interfaces, gastronomy, marketing, and the metaverse.

9.3.1   TRENDS IN INDUSTRY

Industry is one of the most promising fields for the future development of AR. However, industries are still skeptical about using AR in manufacturing processes, mainly because its benefits have not been assessed sufficiently. Noghabaei et al. (2020) investigated the architecture, engineering, and construction industries to discover the benefits of AR technology, and as a result, it was observed to have a significant impact on employees' productivity. The main trends that will be observed in the coming years regarding the use of AR in industries are the following.

- The lack of budget, the lack of understanding of senior managers, and the lack of knowledge of design teams usually limit the implementation of AR. Therefore, industries must learn about the benefits of AR in daily manufacturing processes. Additionally, it is essential that companies not only understand the technology but are also willing to try it.
- Conducting cost and time reduction studies at various stages, from design to maintenance, is essential to convince the industry about the cost–benefit relationship when using AR.
- Efforts must be made to reduce the gap between industry and academia. Both parties must understand that collaborative work will lead to the desired technological development.
- Developers should focus on making AR applications in maintenance, training, and product improvement.

The combination of AR with artificial intelligence is expected to potentialize the creation of smart factories where the quality and production speed increase and the profits are maximized due to low waste generation (Devagiri et al., 2022).

9.3.2 TRENDS IN EDUCATION

AR has been used in education to make teaching–learning more accessible for students and teachers. The interaction, perception, and motivation students experience with AR cannot be offered with traditional teaching methods. In addition, it has been proved that incorporating information and communication technologies (TICs) in educational settings promotes students' desire to learn. Mobile devices are an ideal platform to host educational AR applications, being easy to use and with a high level of interaction and independent operability. Moreover, mobile AR has become popular since students can move freely from one place to another while studying (Irwanto et al., 2022).

However, even though the benefits of AR in educational environments have been proven, there are still opportunities for improvement. Therefore, the trends of AR in the educational field are the following.

- Use non-empirical, qualitative, quantitative, and mixed methods to implement long-term studies to assess the benefits of AR in educational settings.
- Educational AR applications will be designed with teachers to impact learning objectives directly.
- Developers and teachers should carefully analyze the topics in which AR can be used and define the learning materials' requirements and the desired complexity.
- Educational institutions must invest in training teachers in developing technological applications. Unfortunately, the software to rapidly create AR experiences does not offer all the resources to explain complex science topics.
- Mobile devices will continue to be preferred for experimenting with AR in educational settings. However, it is expected that with technological advances,

wearable devices such as AR glasses can lower prices and be introduced as a common tool in classrooms.

Even though it has been stated that AR can be implemented in any educational topic, there are fields in which AR will not have an impact, especially those in which it is unnecessary to observe the phenomenon from different 3D perspectives. Another challenge is to define strategies to motivate teachers to make an extra effort to create new technological content (Kljun et al., 2020). Also, the available technology to display the information in educational settings should be considered (Hernández et al., 2021).

9.3.3 TRENDS IN MULTIMODAL AUGMENTED REALITY

According to Minaee et al. (2022), AR systems are expected to be more advanced soon, using sensors through which visual data will continue to be employed as the base, but the user will also be able to hear, touch, smell, and taste. Azuma (1997) mentioned in his review of AR that "almost all work in AR has focused on the visual sense. Nevertheless, augmentation might apply to all other senses as well." Recent advances in AR have included other sense modalities, such as touch and hearing. However, less progress has been observed in research on smell and taste because, unlike the other senses, the signals are not physical but chemical, making them more difficult to virtualize.

(A) Augmented Reality and Touch

A haptic device can be used in a virtual world to experiment with the sense of touch. Haptics technology employs vibrations and forces feedback to provide tactile information to the user. Therefore, a haptic interface can be designed with cutaneous or kinesthetic feedback. Cutaneous feedback is related to textures and vibrations, while kinesthetic feedback is related to motion and force reactions (Rodríguez et al., 2020). Advances in AR technologies have allowed haptic systems to evolve into multi-sensory feedback, looking to enhance the user experience.

Users expect a certain degree of freedom and portability during an interaction with virtual environments by employing compact devices that enable high-resolution haptic feedback (Yang et al., 2021). It is expected that, in the future, haptic interfaces for AR applications will be able to achieve the following.

- Generate bilateral and multimodal tele-haptic interactions so that users located in different places can share the sense of touch and not only interact visually.
- Develop advanced technology to enable increasingly realistic haptic sensation detection and reproduction through immersive tele-haptic interactions.
- The tactile stimulus is expected to be localized at a specific target point rather than over the entire surface.
- Design sophisticated modulate tangential and frictional force to reproduce textures and multipoint tactile stimulation.

- Inserting haptic interfaces in educational settings. Observing how students interact and learn employing haptic technology could be interesting.

The fusion of AR and haptics will find a niche opportunity in medicine. For example, in the teleoperation area, the fusion will offer information about the direction and distance to the target, contact with the object, and environmental constraints (Lin et al., 2022).

(B)   Augmented Reality and Hearing

Virtual sound can be perceived by augmenting the user's real auditory environment with augmented audio. In augmented audio, the sound is not actually on site, but sensors in the user's headphones can be used to calculate the distance to the source of the virtual sound. AR audio has been little explored in research works (Nagele et al., 2021).

AR audio synthesizes virtual spatialized sounds and favors interaction with the user by having control over the experience, providing information, or directing attention and action. On a museum tour, the user can experience the augmented audio that narrates relevant aspects of the works of art (Yang et al., 2022). Some of the expected trends around augmented audio technology are the following.

- Make transparent and realistic sounds to create an environment where the user cannot differentiate a virtual from a real sound.
- Design AR audio applications to support people with visual impairments or in environments requiring more information while working with heavy machinery.
- Generate immersive and locating experiences using virtual sounds to recreate distances, spectral cues, and directions.
- Develop methods to capture the head-related transfer function (HRTF) adequately. HRTF describes how an ear receives sound from a sound source.
- Design robust tracking algorithms to spatialize sounds appropriately and update acoustics according to user movements.

Spatial audio is an effect a user can experience through headphones or speakers that makes it seem like sound comes from three dimensions. Devices such as the Apple AirPods Pro can produce accurate spatialization of virtual sounds using the integrated module for head tracking (Gupta et al., 2022).

(C)   Augmented Reality and Smell

The chemical sense of smell (aroma) is responsible for detecting and processing odors and is more sensitive than other senses. Humans can discriminate more than 1 trillion olfactory stimuli. Therefore, the smell sense can enrich our experiences in the world around us. Smelling a scent can change a person's mood or transport them to a memory (Wang et al., 2018).

Research in AR regarding the use of smell has generated little interest due to the complexity of producing odors. The first attempt to introduce odors in a virtual environment was presented in the Sensorama simulator, which presented images, sound, vibrations, and the "smells of the city" (Heilig, 1962). Generating digital scents involves combining software engineering, industrial design, electrical engineering, and sensory studies (Kerruish, 2019). Shortly, it is expected that odor-based AR will be able to address the following.

- Convert AR smell research from the exploratory stage to real application in many settings.
- Conduct studies regarding the benefits that could be obtained from implementing olfactory-based AR. One of the benefits could be found in the area of aromatherapy.
- It is expected that, in a few years, the manufacturing of electronic noses will become cheaper and could be converted into an industry with huge gains.
- Through understanding the olfactory space, define a way to code odors from primary scents similarly to how it is done with colors.
- Establish a communication protocol for data exchange between the electronic nose and the AR device. In addition, a way must be designed to understand the data and make them understandable for the user.

Understanding the sense of smell is a scientific challenge that can be divided into two parts: (i) how to track, differentiate, and recognize an odor, and (ii) how to recreate and emit an odor (Erkoyuncu & Khan, 2020). It is expected that thanks to the possibility of smelling the environment, AR experiences will become more immersive.

(D) Augmented Reality and Taste

Undoubtedly, the chemical sense of taste is the least studied human sense in the AR field and the hardest to stimulate digitally. According to Kerruish (2019), taste is the stimulation of taste buds on the human tongue that only gives rise to the five basic tastes of sweet, bitter, salty, sour, and umami. The sense of smell generates approximately 80–90% of the sensations perceived while tasting something. Therefore, the information transduced by the olfactory receptors is referred to the mouth, generating taste sensation (Spence et al., 2017).

Human taste is perceived by combining visual perception, sound, and smell. Therefore, taste, also known as gustation, can be augmented by applying low-amplitude electrical signals or thermal pulses on the human tongue employing tools such as spoons or beverage bottles (Duggal et al., 2022). The trends regarding AR and the olfactory sense are the following.

- Build devices that can be introduced in the market, such as electronic tongues or buccal taste augmentation systems, that do not cause discomfort to the human tongue.
- Virtually augment the sense of taste without involving the other human senses to avoid distractions in the experience.

- Generate flavors that, although synthetic, can be perceived by humans as natural.
- Conduct studies to establish the true values that users would have when using augmented taste, for example, to help users eat only healthy food.

Even when progress is made in augmenting taste, it appears that these techniques will take many years to be implemented on mobile devices. However, it is expected that advances in gustatory augmentation will directly impact the field of digital gastronomy.

9.3.4 TRENDS IN AUGMENTED REALITY AND GASTRONOMY

AR technology has transcended unimaginable disciplines, such as gastronomy, which studies the relationship between food and culture. Digital gastronomy is an emerging field that studies how technologies can be integrated into the kitchen and transform culture, economy, and food preparation (Zoran et al., 2021).

Hotels and restaurants continually look for ways to differentiate themselves from their competition. To achieve this, they try to understand the needs of their consumers and create a way to satisfy them. According to Çöl et al. (2023), there has been increasing interest in adapting AR technology in gastronomy. A menu based on AR can generate an immersive and pleasant customer experience that makes that restaurant unforgettable (Calderón et al., 2023). Gastronomic augmented reality will deal with the following trends.

- Restaurants will use AR to present information about a dish, describe the portions on a plate, or explain the nutritional value of food.
- AR gastronomy must design applications that continually present innovations to prevent diners from no longer considering them attractive.
- Conduct qualitative and quantitative usability, motivation, and satisfaction studies to assess the technology's influence on purchasing decisions.
- Include the senses of taste and smell in the traditional AR experiences; this could be conducted by exploiting chefs' knowledge.
- Employ AR prototypes to perform customized dietary experiences.

AR can also translate or describe menus with a voice and as a marketing tool to attract new customers (Deliyannis et al., 2022). However, it must be clear that virtual experiences will be directed toward diners, so chefs will continue to maintain their artisanal cooking style.

9.3.5 TRENDS IN AUGMENTED REALITY AND MARKETING

Technologies can help develop new marketing strategies to revolutionize how products and services are sold. The process of using technology such as an electronic device or the Internet to communicate, promote, or sell the value of a product or service is known as digital marketing (Grewal et al., 2020). AR is one of the technological tools that has begun to be used successfully in digital marketing. AR allows companies to promote a product or service through interactive experiences that mix the real world and computer-generated representations of the product (Chylinski et al., 2020).

AR will change the rules for digital marketing because it captures users' attention for longer, improves the shopping experience, provides a more aesthetic image of the product or service, and increases the chances for purchase (Ng & Ramasamy, 2018). The main trends that will be observed in the area of AR marketing in the coming years are the following.

- Use of multimodal AR to change the traditional way of displaying products. It is expected that the user can feel, hear, and see the product as if they had it in their hands.
- Use AR to develop long-term marketing campaigns to study buyer behavior and propose strategies to keep them captive.
- Bring the purchasing process to the next level by generating AR virtual try-on. A virtual try-on allows consumers to observe how shoes, clothes, makeup, or jewelry, look on by using a mobile device's camera.
- Companies will begin to see the success of their competition and become convinced of AR's benefits for digital marketing.

In a few years, shoppers will prefer stores offering AR experiences regardless of whether it means paying more for the product. Moreover, AR strategies for digital marketing will benefit from advances in metaverse development.

9.3.6 TRENDS IN AUGMENTED REALITY AND METAVERSE

Metaverse can be understood as an Internet application that mixes a variety of technologies (AR, digital twins, artificial intelligence, blockchain) to generate a 3D environment where people can access remotely from various locations for social interaction using avatars (Wang et al., 2023). Metaverse combines two words: meta meaning occurring later and universe meaning everything. Moreover, AR is one of the key technologies directly related to the metaverse development.

Education and healthcare are two fields in which the metaverse has the most impact. The metaverse and AR in educational settings offer students a space for social communication in which they are free to create and share knowledge (Kye et al., 2021). On the other hand, AR and metaverse support students learning about the human body in the healthcare field (Bhugaonkar et al., 2022). Therefore, the trends that will be observed are the following.

- Promotion of the use and advantages of the metaverse so that it can move from the experimentation stage to the application stage in different environments.
- Due to their innovative characteristics, the metaverse and AR are expected to help generate teaching–learning strategies that increase student motivation.
- Design a secure platform using security and privacy protocols to prevent misuse of personal data and not put users at risk.

- Promote cooperative work among users through project solutions employing educational metaverse.
- Build hands-free wearables that combine, at least, the sense of sight and touch.

It is expected that with the increase in the computational power of mobile devices, users will be able to try metaverse and AR experiences with excellent visual quality and multimodal interaction.

REFERENCES

Alalwan, N., Cheng, L., Al-Samarraie, Y., Ibrahim, A., & Sarsam, S. (2020). Challenges and Prospects of Virtual Reality and Augmented Reality Utilization among Primary School Teachers: A Developing Country Perspective. *Studies in Educational Evaluation, 66*, 1–12. https://doi.org/https://doi.org/10.1016/j.stueduc.2020.100876

Alismail, A., Altulaihan, E., Rahman, M., & Sufian, A. (2023). A Systematic Literature Review on Cybersecurity Threats of Virtual Reality (VR) and Augmented Reality (AR). In Jacob, I., Kolandapalayam, S., & Izonin, I. (Eds.), *Proceedings of the Data Intelligence and Cognitive Informatics* (pp. 761–774). Springer Nature Singapore.

Altinpulluk, H. (2019). Determining the Trends of Using Augmented Reality in Education Between 2006–2016. *Education and Information Technologies, 24*(2), 1089–1114. https://doi.org/10.1007/s10639-018-9806-3

Alzahrani, N. (2020). Augmented Reality: A Systematic Review of Its Benefits and Challenges in E-learning Contexts. *Applied Sciences, 10*(16), 1–21. https://doi.org/10.3390/app10165660

Angrisani, L., Arpaia, P., Esposito, A., & Moccaldi, N. (2020). A Wearable Brain–Computer Interface Instrument for Augmented Reality-Based Inspection in Industry 4.0. *IEEE Transactions on Instrumentation and Measurement, 69*(4), 1530–1539. https://doi.org/10.1109/TIM.2019.2914712

Ateya, A., Muthanna, A., Koucheryavy, A., Maleh, Y., & El-Latif, A. (2023). Energy Efficient Offloading Scheme for MEC-Based Augmented Reality System. *Cluster Computing, 26*(1), 789–806. https://doi.org/10.1007/s10586-022-03914-7

Azuma, R. (1997). A Survey of Augmented Reality. *Presence: Teleoperators and Virtual Environments, 6*(4), 355–385. https://doi.org/10.1162/pres.1997.6.4.355

Bacca, J., Baldiris, S., Fabregat, R., Graf, S., & Kinshuk. (2014). Augmented Reality Trends in Education: A Systematic Review of Research and Applications. *Journal of Educational Technology & Society, 17*(4), 133–149. http://www.jstor.org/stable/jeductechsoci.17.4.133

Bhugaonkar, K., Bhugaonkar, R., & Masne, N. (2022). The Trend of Metaverse and Augmented & Virtual Reality Extending to the Healthcare System. *Cureus, 14*(9), 1–7. https://doi.org/10.7759/cureus.29071

Billinghurst, M. (2021). Grand Challenges for Augmented Reality. *Frontiers in Virtual Reality, 2*, 1–4. https://doi.org/10.3389/frvir.2021.578080

Billinghurst, M., Kato, H., & Poupyrev, I. (2008). Tangible Augmented Reality. *Proceedings of the ACM Siggraph Asia, 7*(2), 1–10.

Billinghurst, M., Piumsomboon, T., & Bai, H. (2014). Hands in Space: Gesture Interaction with Augmented-Reality Interfaces. *IEEE Computer Graphics and Applications, 34*(1), 77–80. https://doi.org/10.1109/MCG.2014.8

Brown, C., Hicks, J., Rinaudo, C., & Burch, R. (2023). The Use of Augmented Reality and Virtual Reality in Ergonomic Applications for Education, Aviation, and Maintenance. *Ergonomics in Design, 31*(4), 23–31. https://doi.org/10.1177/10648046211003469

Calderón, V., Carrasco, M., & Rossi, C. (2023). The Intention of Consumers to Use Augmented Reality Apps in Gastronomy—Case of Málaga. *Current Issues in Tourism, 26*(9), 1446–1462. https://doi.org/10.1080/13683500.2022.2056002

Chen, X., Huang, X., Wang, Y., & Gao, X. (2020). Combination of Augmented Reality Based Brain- Computer Interface and Computer Vision for High-Level Control of a Robotic Arm. *IEEE Transactions on Neural Systems and Rehabilitation Engineering, 28*(12), 3140–3147. https://doi.org/10.1109/TNSRE.2020.3038209

Chylinski, M., Heller, J., Hilken, T., Keeling, D., Mahr, D., & de Ruyter, K. (2020). Augmented Reality Marketing: A Technology-Enabled Approach to Situated Customer Experience. *Australasian Marketing Journal, 28*(4), 374–384. https://doi.org/10.1016/j.ausmj.2020.04.004

Çöl, B., İmre, M., & Yıkmış, S. (2023). Virtual Reality and Augmented Reality Technologies in Gastronomy: A Review. *EFood, 4*(3), 1–16. https://doi.org/10.1002/efd2.84

Danielsson, O., Holm, M., & Syberfeldt, A. (2020). Augmented Reality Smart Glasses in Industrial Assembly: Current Status and Future Challenges. *Journal of Industrial Information Integration, 20*, 1–15. https://doi.org/https://doi.org/10.1016/j.jii.2020.100175

Datcu, D., Lukosch, E., & Frances, B. (2015). On the Usability and Effectiveness of Different Interaction Types in Augmented Reality. *International Journal of Human–Computer Interaction, 31*(3), 193–209. https://doi.org/10.1080/10447318.2014.994193

Deliyannis, I., Poulimenou, S., Kaimara, P., & Laboura, S. (2022). BRENDA Digital Tours: Designing a Gamified Augmented Reality Application to Encourage Gastronomy Tourism and Local Food Exploration. In Vujicic, M., Kasim, A., Kostopoulou, S., Chica, J., Aslam, M. (eds) *Cultural Sustainable Tourism* (pp. 101–109). https://doi.org/10.1007/978-3-031-07819-4_9

Devagiri, J., Paheding, S., Niyaz, Q., Yang, X., & Smith, S. (2022). Augmented Reality and Artificial Intelligence in industry: Trends, Tools, and Future Challenges. *Expert Systems with Applications, 207*, 1–40. https://doi.org/https://doi.org/10.1016/j.eswa.2022.118002

Dirin, A., & Laine, T. (2018). User Experience in Mobile Augmented Reality: Emotions, Challenges, Opportunities and Best Practices. *Computers, 7*(2), 1–18. https://doi.org/10.3390/computers7020033

Doerner, R., & Horst, R. (2022). Overcoming Challenges when Teaching Hands-On Courses about Virtual Reality and Augmented Reality: Methods, Techniques and Best Practice. *Graphics and Visual Computing, 6*, 1–11. https://doi.org/https://doi.org/10.1016/j.gvc.2021.200037

Duggal, A., Singh, R., Gehlot, A., Rashid, M., Alshamrani, S., & AlGhamdi, A. (2022). Digital Taste in Mulsemedia Augmented Reality: Perspective on Developments and Challenges. *Electronics, 11*(9), 1315. https://doi.org/10.3390/electronics11091315

Erkoyuncu, J., & Khan, S. (2020). Olfactory-Based Augmented Reality Support for Industrial Maintenance. *IEEE Access, 8*, 30306–30321. https://doi.org/10.1109/ACCESS.2020.2970220

García, I., Portalés, C., Gimeno, J., & Casas, S. (2020). A Collaborative Augmented Reality Annotation Tool for the Inspection of Prefabricated Buildings. *Multimedia Tools and Applications, 79*(9), 6483–6501. https://doi.org/10.1007/s11042-019-08419-x

Garzón, J. (2021). An Overview of Twenty-Five Years of Augmented Reality in Education. *Multimodal Technologies and Interaction, 5*(7), 1–14. https://doi.org/10.3390/mti5070037

Grewal, D., Hulland, J., Kopalle, P., & Karahanna, E. (2020). The Future of Technology and Marketing: A Multidisciplinary Perspective. *Journal of the Academy of Marketing Science, 48*(1), 1–8. https://doi.org/10.1007/s11747-019-00711-4

Gupta, R., He, J., Ranjan, R., Gan, W., Klein, F., Schneiderwind, C., Neidhardt, A., Brandenburg, K., & Välimäki, V. (2022). Augmented/Mixed Reality Audio for Hearables: Sensing, Control, and Rendering. *IEEE Signal Processing Magazine, 39*(3), 63–89. https://doi.org/10.1109/MSP.2021.3110108

Heilig, M. (1962). *Sensorama Simulator* (Patent US3050870). https://patents.google.com/pat ent/US3050870A/en

Hernández, L., López, J., Tovar, M., Vergara, O., & Cruz, V. (2021). Effects of Using Mobile Augmented Reality for Simple Interest Computation in a Financial Mathematics Course. *PeerJ Computer Science, 7:e618*(1), 1–33.

Huang, B., Lin, C., & Lee, C. (2012). Mobile Augmented Reality Based on Cloud Computing. *Proceedings of the Anti-Counterfeiting, Security, and Identification*, 1–5. https://doi.org/ 10.1109/ICASID.2012.6325354

Hussain, M., Park, J., & Kim, H. (2023). Augmented Reality Sickness Questionnaire (ARSQ): A refined Questionnaire for Augmented Reality Environment. *International Journal of Industrial Ergonomics*, *97*, 1–10. https://doi.org/https://doi.org/10.1016/ j.ergon.2023.103495

Irwanto, I., Dianawati, R., & Lukman, I. (2022). Trends of Augmented Reality Applications in Science Education: A Systematic Review from 2007 to 2022. *International Journal of Emerging Technologies in Learning (IJET)*, *17*(13), 157–175. https://doi.org/10.3991/ ijet.v17i13.30587

Kaufeld, M., Mundt, M., Forst, S., & Hecht, H. (2022). Optical See-Through Augmented Reality can Induce Severe Motion Sickness. *Displays*, *74*, 102283. https://doi.org/ https://doi.org/10.1016/j.displa.2022.102283

Kerruish, E. (2019). Arranging Sensations: Smell and Taste in Augmented and Virtual Reality. *Senses and Society*, *14*(1), 31–45. https://doi.org/10.1080/17458927.2018.1556952

Kim, J., Ari, H., Madasu, C., & Hwang, J. (2020). Evaluation of the Biomechanical Stress in the Neck and Shoulders During Augmented Reality Interactions. *Applied Ergonomics*, *88*, 1–9. https://doi.org/https://doi.org/10.1016/j.apergo.2020.103175

Kim, J., Laine, T., & Åhlund, C. (2021). Multimodal Interaction Systems Based on Internet of Things and Augmented Reality: A Systematic Literature Review. *Applied Sciences*, *11*(4), 1–33. https://doi.org/10.3390/app11041738

Kim, K., Billinghurst, M., Bruder, G., Duh, H., & Welch, G. (2018). Revisiting Trends in Augmented Reality Research: A Review of the 2nd Decade of ISMAR (2008–2017). *IEEE Transactions on Visualization and Computer Graphics*, *24*(11), 2947–2962. https://doi.org/10.1109/TVCG.2018.2868591

Kljun, M., Geroimenko, V., & Čopič, K. (2020). Augmented Reality in Education: Current Status and Advancement of the Field. In Geroimenko, V. (Ed.), *Augmented Reality in Education: A New Technology for Teaching and Learning* (pp. 3–21). Springer International Publishing. https://doi.org/10.1007/978-3-030-42156-4_1

Kye, B., Han, N., Kim, E., Park, Y., & Jo, S. (2021). Educational Applications of Metaverse: Possibilities and Limitations. *Journal of Educational Evaluation for Health Professions*, *18*(32), 1–13. https://doi.org/10.3352/jeehp.2021.18.32

Lampropoulos, G., Keramopoulos, E., & Diamantaras, K. (2020). Enhancing the Functionality of Augmented Reality using Deep Learning, Semantic web and Knowledge Graphs: A Review. *Visual Informatics*, *4*(1), 32–42. https://doi.org/https://doi.org/10.1016/j.vis inf.2020.01.001

Lanham, M. (2018). *Learn ARCore-Fundamentals of Google ARCore: Learn to Build Augmented Reality Apps for Android, Unity, and the Web with Google ARCore 1.0*. Packt Publishing Ltd.

Lin, T., Krishnan, A., & Li, Z. (2022). Comparison of Haptic and Augmented Reality Visual Cues for Assisting Tele-manipulation. Proceedings of the International Conference on Robotics and Automation (ICRA), 9309–9316. https://doi.org/10.1109/ICRA46 639.2022.9811669

Marques, B., Silva, S., Alves, J., Araújo, T., Dias, P., & Santos, B. (2022). A Conceptual Model and Taxonomy for Collaborative Augmented Reality. *IEEE Transactions on Visualization and Computer Graphics*, *28*(12), 5113–5133. https://doi.org/10.1109/ TVCG.2021.3101545

Masood, T., & Egger, J. (2019). Augmented Reality in Support of Industry 4.0—Implementation Challenges and Success Factors. *Robotics and Computer-Integrated Manufacturing*, *58*, 181–195. https://doi.org/https://doi.org/10.1016/j.rcim.2019.02.003

Minaee, S., Liang, X., & Yan, S. (2022). *Modern Augmented Reality: Applications, Trends, and Future Directions*, pp. 1–24, *arXiv preprint arXiv:2202.09450*. doi:10.48550/ arXiv.2202.09450

Muñoz, L., Miró, L., & Domínguez, M. (2020). Augmented and Virtual Reality Evolution and Future Tendency. *Applied Sciences*, *10*(1), 1–23. https://doi.org/10.3390/app10010322

Nagele, A., Bauer, V., Healey, P., Reiss, J., Cooke, H., Cowlishaw, T., Baume, C., & Pike, C. (2021). Interactive Audio Augmented Reality in Participatory Performance. *Frontiers in Virtual Reality*, *1*, 1–14. https://doi.org/10.3389/frvir.2020.610320

Neely, E. (2019). Augmented Reality, Augmented Ethics: Who Has the Right to Augment a Particular Physical Space? *Ethics and Information Technology*, *21*(1), 11–18. https:// doi.org/10.1007/s10676-018-9484-2

Ng, C., & Ramasamy, C. (2018). Augmented Reality Marketing in Malaysia—Future Scenarios. *Social Sciences*, *7*(11), 1–15. https://doi.org/10.3390/socsci7110224

Noghabaei, M., Heydarian, A., Balali, V., & Han, K. (2020). Trend analysis on adoption of virtual and augmented reality in the architecture, engineering, and construction industry. *Data*, *5*(1), 1–18. doi:10.3390/data5010026

Qiao, X., Ren, P., Nan, G., Liu, L., Dustdar, S., & Chen, J. (2019). Mobile Web Augmented Reality in 5G and Beyond: Challenges, Opportunities, and Future Directions. *China Communications*, *16*(9), 141–154. https://doi.org/10.23919/JCC.2019.09.010

Rejeb, A., Keogh, J., Leong, G., & Treiblmaier, H. (2021). Potentials and Challenges of Augmented Reality Smart Glasses in Logistics and Supply Chain Management: A Systematic Literature Review. *International Journal of Production Research*, *59*(12), 3747–3776. https://doi.org/10.1080/00207543.2021.1876942

Rodríguez, A., Nandayapa, M., Vergara, O., & García, F. (2020). Haptic Augmentation Towards a Smart Learning Environment: The Haptic Lever Design. *IEEE Access*, *8*, 78467–78481. https://doi.org/10.1109/ACCESS.2020.2990172

Roesner, F., Kohno, T., & Molnar, D. (2014). Security and Privacy for Augmented Reality Systems. *Communications of the ACM*, *57*(4), 88–96. https://doi.org/10.1145/2580 723.2580730

Sabelman, E., & Lam, R. (2015). The Real-Life Dangers of Augmented Reality. *IEEE Spectrum*, *52*(7), 48–53. https://doi.org/10.1109/MSPEC.2015.7131695

Spence, C., Obrist, M., Velasco, C., & Ranasinghe, N. (2017). Digitizing the Chemical Senses: Possibilities & Pitfalls. *International Journal of Human-Computer Studies*, *107*, 62–74. https://doi.org/10.1016/j.ijhcs.2017.06.003

Syiem, B., Kelly, R., Goncalves, J., Velloso, E., & Dingler, T. (2021). Impact of Task on Attentional Tunneling in Handheld Augmented Reality. Proceedings of the 2021 CHI Conference on Human Factors in Computing Systems, 1–14. https://doi.org/10.1145/ 3411764.3445580

Villanueva, A., Zhu, Z., Liu, Z., Wang, F., Chidambaram, S., & Ramani, K. (2022). ColabAR: A Toolkit for Remote Collaboration in Tangible Augmented Reality Laboratories. *Proceedings of the ACM on Humand-Computer Interaction*, *6*(1), 1–22. https://doi.org/10.1145/3512928

Wang, J., Erkoyuncu, J., & Roy, R. (2018). A Conceptual Design for Smell Based Augmented Reality: Case Study in Maintenance Diagnosis. *Procedia CIRP*, *78*, 109–114. https://doi.org/https://doi.org/10.1016/j.procir.2018.09.067

Wang, W. (2018). Understanding Augmented Reality and ARKit. In Beginning ARKit for iPhone and iPad: Augmented Reality App Development for iOS (pp. 1–17). Apress. https://doi.org/10.1007/978-1-4842-4102-8_1

Wang, Y., Su, Z., Zhang, N., Xing, R., Liu, D., Luan, T. H., & Shen, X. (2023). A Survey on Metaverse: Fundamentals, Security, and Privacy. *IEEE Communications Surveys & Tutorials*, *25*(1), 319–352. https://doi.org/10.1109/COMST.2022.3202047

Yang, J., Barde, A., & Billinghurst, M. (2022). Audio Augmented Reality: A Systematic Review of Technologies, Applications, and Future Research Directions. *Journal of the Audio Engineering Society*, *70*(10), 788–809. https://doi.org/10.17743/jaes.2022.0048

Yang, T., Kim, J., Jin, H., Gil, H., Koo, J., & Kim, H. (2021). Recent Advances and Opportunities of Active Materials for Haptic Technologies in Virtual and Augmented Reality. *Advanced Functional Materials*, *31*(39), 1–30. https://doi.org/10.1002/adfm.202008831

Zhan, T., Yin, K., Xiong, J., He, Z., & Wu, S. (2020). Perspective Augmented Reality and Virtual Reality Displays: Perspectives and Challenges. *IScience*, *23*(8), 1–13. https://doi.org/10.1016/j.isci

Zhang, X., Vadodaria, H., Li, N., Kang, K., & Liu, Y. (2020). A Smartphone Thermal Temperature Analysis for Virtual and Augmented Reality. *Proceedings of the IEEE International Conference on Artificial Intelligence and Virtual Reality (AIVR)*, 301–306. https://doi.org/10.1109/AIVR50618.2020.00061

Zoran, A., Gonzalez, E. A., Mizrahi, A. B., & Lachnish, A. "Zoonder." (2021). Cooking with Computers: The Vision of Digital Gastronomy. In Galanakis, C. (ed.) *Gastronomy and Food Science* (pp. 35–53). Elsevier. https://doi.org/10.1016/B978-0-12-820 057-5.00003-0

10 Concluding Remarks

10.1 SUMMARY

AR is about artificially improving a user's perspective of a real environment. The origins of AR date back to the 1960s when Sutherland (1968) presented proposals about how computers would become a window into virtual worlds. The first AR applications used heavy and expensive head-mounted displays, and due to the system configuration, user mobility was almost non-existent. However, since then, there have been expectations of how technological development could help the consolidation of AR.

Nowadays, technological development has opened the door to a new era of AR applications where mobility and computational power are the main features. Moreover, multimodal AR applications that involve more than one sense are beginning to be built. The history of AR continues to be written daily. Even though it was asserted that AR is in its early stages, it will not be many years before we see its consolidation. Therefore, AR should no longer be considered an emerging technology but a technology entering its maturity stage.

In the book's first chapter, the readers learn about the definition and different types of AR. In addition, a historical review of the most important events related to the development of AR was provided. It was also explained that although the terms AR and VR differ, they are usually used similarly. The former enriches real-world scenes by adding computer-generated virtual objects. Meanwhile, the latter shows the user a completely immersive scenario of a scene in which all components are virtual.

In the second chapter, the reader was informed about seven AR applications in education, four in medicine, and four in industry. All the applications were designed by bachelor's, master's, and doctoral students of the authors of this book. In addition, it was explained that the two fundamental skills to realize AR applications are computer vision and programming.

Chapter 3 explained the set of steps that digital image processing uses to develop an AR application. We explained how to acquire, process, segment, characterize, and recognize images. In addition, the procedures for camera calibration, tracking, and registration techniques were explained.

The fourth chapter presented an explanation of the technological devices to display AR. The devices were divided into head-mounted and non-head-mounted. Regarding

DOI: 10.1201/9781003435198-10

193

HMDs, the non-see-through, optical see-through, and video see-through devices were presented. On the other hand, handheld displays and spatial displays were explained in the non-head-mounted category.

In Chapter 5, a cycle for developing AR applications was proposed. The developer is assumed to have already completed the requirements analysis process. The cycle includes the stages of development of virtual models and markers, programming, deployment, and user experience. In addition, it was explained to the reader how to realize the first AR application with Meta Spark Studio step by step.

The sixth chapter explained the different instruments to evaluate an AR application's usability, acceptance, motivation, immersion, and quality. An explanation of each instrument and instructions to apply it were shown.

Chapters 7 and 8 provided two different case studies of AR implementation. The example in Chapter 7 was called ARGeo, which supports students in learning geography, mainly to learn about the Earth's layers. Moreover, motivation, immersion, quality, technology acceptance, and student achievement were assessed.

Chapter 8 described a markerless application to support the assembly process of an educational dragster toy. The usability of the tool and the time to complete the assembly task were assessed.

Chapter 9 discussed the technological, usability, and social challenges of AR. Also, it presented trends in the fields of industry, education, multimodal AR, gastronomy, marketing, and the metaverse.

Finally, this chapter summarizes the book, proposes a framework to develop AR applications, explains the importance of writing papers, standards, and patents, and shares some lessons learned by the book's authors.

What is a fact is that AR has come into our lives and is here to stay. Several factors will help the consolidation of AR, such as the increase in storage capacity, the everyday use of mobile devices, the increase in the computational power of mobile devices, the Internet, cloud computing, and, in general, the habit of people of using technology daily. Moreover, a framework to develop AR apps, the writing of scientific papers, the standardization of AR, and an increase in patenting culture will also help AR consolidation. These issues are discussed in the following subsections.

10.2 PROPOSAL OF A FRAMEWORK TO DEVELOP AR APPLICATIONS

Several novel works have assessed the effectiveness of using AR, mainly in educational settings (Buchner & Kerres, 2023; Chang et al., 2022; López et al., 2023). However, to support the integration of AR into daily life, there are still open challenges, and one of them is to design a robust, configurable, extensible, and economical framework on which programmers can rely to begin standardizing application development.

Frameworks such as ComposAR (Seichter et al., 2008), Studierstube (Schmalstieg et al., 2002), ARTiFiCe (Mossel et al., 2012), and Designers Augmented Reality Toolkit (DART) (MacIntyre et al., 2005) have been developed; unfortunately, none have gained general acceptance.

Despite the variety and differences between applications, most AR systems share a common basic architectural structure (MacWilliams et al., 2004). All AR systems are interactive, and their main functionality is tracking the position of a marker, mixing virtual objects with real scenes, and processing and reacting to the interaction made by the user. Therefore, we can find fundamental subsystems and components in most AR applications. Figure 10.1 shows the proposed simple, flexible, and extensible framework based on computer vision techniques and reusable components to develop mobile AR applications (Barraza et al., 2015). The explanation of each module is presented in the following subsections.

10.2.1   AR APPLICATION

In the first stage, the developer must collect the requirements from the client. The developer must use all their experience to determine if the process the client needs can be visualized with AR. If AR is the ideal tool, then the number of markers and associated 3D models must be determined. The reader should remember that AR can be performed with and without markers. The design must be carried out for the case with markers; for the case without markers, it must be determined which real objects within the scene will trigger the AR experience.

The 3D models must be designed carefully to make them as real as possible. However, it should be considered that, due to the characteristics of mobile devices, it is not always possible to display 3D models on the screen optimally. Once the markers and 3D models are made, the operating system and the device in which the AR application will run must be determined, and after that, the development of the prototype starts by building four subsystems.

10.2.2   RENDERING SUBSYSTEM

The rendering subsystem is the framework's core and is based on the Unity3D game engine. Unity3D is employed for rendering, generating transformation models, handling lighting, shading, and special effects, making physical simulations, and detecting collisions (Linowes & Babilinski, 2017). Therefore, the rendering subsystem allows displaying the augmented scene to the user mixed with virtual and real-world elements.

Unity3D is one of the preferred engines for AR developers because it allows exporting the final development to different platforms, including Windows, Mac OS, iOS, and Android. Moreover, Unity3D is flexible, extendable, ideal for handling scenes, and can extend its functionality through direct programming or using extensions from independent developers.

Every element inside a Unity3D scene is derived from the GameObject base class. The class is a hierarchical container of objects called "components." The components are responsible for the system functionality. Therefore, because every element derives from the GameObject class, the common properties and methods, such as the renderer, the animator, the collision detector, and the transformer, are inherited. Moreover, the

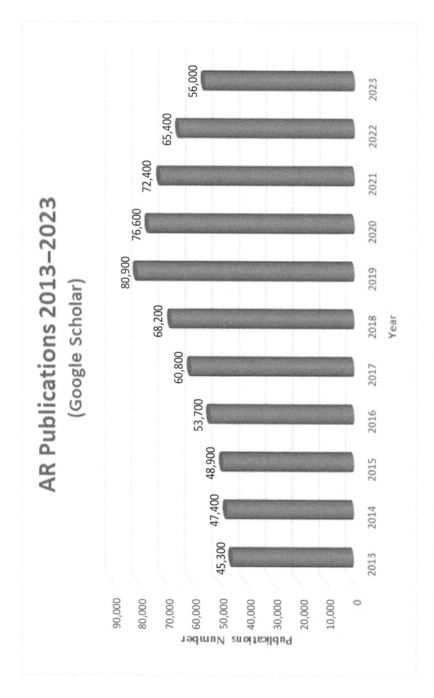

FIGURE 10.1   Framework to develop mobile AR apps (Barraza, 2015).

C# language programming can be employed to develop personalized classes that allow an object to respond to user interaction through the interaction subsystem.

10.2.3 TRACKING SUBSYSTEM

The tracking subsystem executes the marker detection, processing, thresholding, filtering, and pose estimation. The information collected from these tasks is sent to the rendering subsystem to generate the final scene displayed to the user. Therefore, the tracking subsystem is key to developing the AR application. The jitters and fluctuations induced by the tracking must be kept to a minimum without sacrificing the response time to maintain the illusion that the real and virtual worlds coexist as one.

The main functionality of the tracking subsystem is based on the Vuforia Software Developer Kit (SDK) (Simonetti & Paredes, 2013). Vuforia is incorporated into Unity3D as an extension. Therefore, the source code from the rendering and tracking can be called from the TrackableBehaviour class, which is derived from the base class MonoBehaviour to facilitate communication and data interchange.

The tracking subsystem includes the AR camera and the associated scripts to ensure that the tracking and acquisition finalization are conducted in each frame acquired. Hence, Vuforia transforms the frames acquired into OpenGL to conduct the rendering and tracking operations. Object detection and tracking are conducted for each frame acquired, and the information obtained is stored in a state object. The rendering subsystem employs the state object to ensure correct positioning and representation of the 3D object in the video feed of the real world. The tracking–update loop is executed for every frame to be processed.

10.2.4 INTERACTION SUBSYSTEM

The interaction subsystem collects and processes any input action the user performs. In the case of mobile AR, it is proposed that user interaction be carried out through a tactile graphical interface because most mobile device users are comfortable with this type of interaction.

The graphical user interface must always be visible and ready to transmit information to the rendering subsystem when user interaction with the device is detected. The transmitted values modify the transformation and rendering properties of the virtual objects (pose estimation). Therefore, a custom class is created and associated with the AR camera object in this subsystem.

When a screen click is registered, a series of events are triggered to evaluate whether an element of the graphical interface was touched. If the evaluation is positive, it is checked if a valid marker appears in the scene. Finally, if there is a valid marker in the scene, the calculations are carried out, the information is transmitted to the rendering subsystem, and, if necessary, the display of the virtual object is updated. The C# classes of this subsystem draw the user controls on the device screen and perform the operations when an interaction succeeds.

10.2.5 CONTEXT AND WORLD MODEL SUBSYSTEM

The context and world model subsystems are used together to generate the visual scene editor. The context subsystem stores and provides relevant contextual information to the entire framework. The information can be static or generated in real time when the application runs. A custom class associated with an empty Unity3D object and the whiteboard pattern ensures that all objects can read and write the necessary information. Alternatively, a pattern where objects communicate directly with each other to exchange data can be used.

The world model subsystem stores and provides access to the digital representation of the real world. The representation includes the reference marker patterns, data about interest points, and the 3D models used for augmentations. The world model subsystem links the tracking library classes and the scene editor.

10.2.6 COMPUTING PLATFORMS

The computing platforms include all the hardware elements used to develop the AR App. Knowing what central processing unit (CPU) will be employed is essential in this stage. Also, it is important to know the set of sensors included in the mobile device.

Another important consideration is the operating system in which the AR app will be implemented. For the case of mobile devices, two main options are available: (i) Android operating system and ii) iPhone operating system (iOS).

10.2.7 INPUT/OUTPUT DEVICE

Regarding mobile AR, three different devices can be employed to conduct input and output operations, such as: (i) mobile phones, (ii) tablet computers, and (iii) head-mounted displays (HMDs). All three devices include touch inputs and outputs through a screen. The study cases shown in Chapters 7 and 8 were developed following the framework proposed in this subsection.

10.3 AR PAPERS

Since Caudell coined the term AR (Caudell & Mizell, 1992), many researchers have shown interest in conducting studies and presenting innovative applications in this fascinating area. Scientific writing is a fundamental part of any research process because when an article is published in a journal or at a conference proceedings, a contribution is made to the field of knowledge (Graham et al., 2013). Furthermore, the publication of a paper implies that experts of a particular discipline domain validated it.

A scientific paper reports the results of an investigation and must follow a standardized structure regarding its content and space (number of pages). The traditional structure of a scientific paper is the following: (i) Title, (ii) Abstract, (iii) Keywords, (iv) Introduction, (v) Theoretical background, (vi) Related work, (vii) Materials and methods, (viii) Experiments and results, (ix) Discussion,

(x) Conclusions, (xi) Further work, (xii) Acknowledgments, (xiii) References, and (xiv) Appendices (Suppe, 1998).

Journal papers have become the most important source for sharing scientific information; without them, science could not exist. Moreover, papers must provide enough information to replicate procedures successfully tested by the authors. Regarding AR, many papers have been written. However, the field in which the most findings have been presented is education. Moreover, the publication of the "Survey of Augmented Reality" by Azuma (1997) marked a before and after in the field of AR.

As can be observed in Figure 10.2 around 675,000 papers have been published addressing the AR topic in the last 10 years (2013–2023). The year 2019 was when most publications were made, and from there, a decline began. This decrease does not mean interest in AR has been lost. It means that more people have become experts on the subject, and evaluating and publishing an article has become more rigorous than ever.

Recent reviews argue that education, medicine, robotics, manufacturing, and entertainment are the fields in which most publications about AR were presented (Arena et al., 2022; Dargan et al., 2023; López et al., 2023). However, as explained in Chapter 9, coming works are expected to focus on developing industry applications and exploiting the benefits of the new technological devices.

10.4   AR STANDARDS

Generating standards is crucial for countries because they determine access to specific segments of a market and the terms of participation in global value chains. A standard is a documented agreement containing precise technical specifications and criteria to be used consistently as rules, guides, or defining characteristics to ensure that materials, products, processes, and services meet their intended purpose (Nadvi & Waltring, 2004).

A standard addresses labor conditions, quality management procedures, health and safety norms, and environmental and social concerns (Giovannucci & Ponte, 2005). The Institute of Electrical and Electronics Engineers (IEEE) has been working hard to develop AR standards. Standards define best practices for obtaining requirements and implementing AR in different settings. The benefits of creating and implementing AR standards are threefold. First, a baseline of interoperability between manufacturers and content publishers will be established. Second, standards will facilitate the development of client applications. Finally, with the content availability and the variety of use cases and devices, the implementation cost will be reduced, and the risk investment will be mitigated (Perey et al., 2011).

Currently, the IEEE P2048 working group is working on developing 12 standards regarding AR and VR. The standards meetings enable all the participants to discuss use cases, requirements, and issues related to AR. Over 200 companies and institutions worldwide are participating in the working group. The participants include device manufacturers, content and service providers, technology developers, government agencies, and other parties relevant to VR/AR and mixed reality (MR). The 12 developed standards focus on the following areas (Yuan, 2018).

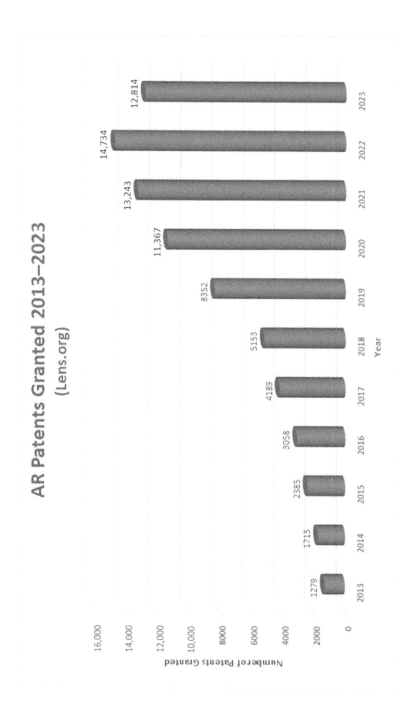

FIGURE 10.2   Graph of the number of AR publications (2013–2023).

1. IEEE P2048.1: Device taxonomy and definitions. Seeks to divide the VR/AR devices into different categories and levels to reduce confusion in devices with similar product names but that perform significantly different functions.
2. IEEE P2048.2: Immersive video taxonomy and quality metrics. Focus on dividing immersive video into different categories and levels to help users choose the right products and promote the industry's health.
3. IEEE P2048.3: Immersive video files and stream formats. Specifies the immersive video file and stream formats supported and facilitates the development of cross-platform content and services.
4. IEEE P2048.4: Person identity. States the requirements and methods for verifying the identity of a person.
5. IEEE P2048.5: Environment safety. Stipulates recommendations for workstation and content consumption environment for VR, AR, MR, and all related devices where a digital overlay might interact with the physical world, potentially impacting users' perception.
6. IEEE P2048.6: Immersive user interface. Specifies the requirement interactions, methods, and requirements for enabling the immersive user interface in VR/AR applications.
7. IEEE P2048.7: Map for virtual objects in the real world. Dictates the systems, methods, requirements, tests, and verifications for AR/MR to assign coordinates and orientations for virtual objects in the real world.
8. IEEE P2048.8: Interoperability between virtual objects and the real world. Governs different categories and levels of interoperability between virtual objects and the real world and specifies the systems and methods that enable these categories and levels.
9. IEEE P2048.9: Immersive audio taxonomy and quality metrics. Specifies the taxonomy and quality metrics for immersive audio.
10. IEEE P2048.10: Immersive audio file and stream formats. Identifies the formats of immersive audio files and streams and the functions and interactions enabled by the formats.
11. IEEE P2048.11: In-vehicle augmented reality. Defines a framework for AR systems that assists vehicle drivers and passengers and makes user interfaces more friendly.
12. IEEE P2048.12: Content ratings and descriptors. Establishes the content ratings and descriptors to protect users' health and safety from risky VR, AR, and MR.

Other available standards are the IEEE 1589-2020 for AR learning experience model, IEEE IC16-004-02 for AR in the oil/gas/electric industry, and IEEE P2048.101 for AR on mobile devices: general requirements for software framework, components, and integration. Therefore, it is expected that soon all the standards will be concluded and help the consolidation and public acceptance of AR and consequently experiment with industry insertion of the technology.

10.5   AR PATENTS

Every day, countries work to achieve economic and social development, and creating innovative technology is one activity that leads to attaining this goal. Therefore, when a technological innovation is generated, it is crucial to protect it; typically, this action is carried out through intellectual property. Intellectual property refers to protecting the product of human intellect in the scientific, literary, artistic, or industrial fields (Bently et al., 2022). There are at least two ways to protect intellectual property: (i) industrial property (patents, utility models, industrial designs, and distinctive signs) and (ii) copyrights (literary, musical, artistic, and photographic works).

A patent is an exclusive temporary right granted over an invention that allows its owner to decide whether or not third parties can use it. Moreover, a patent is a product or process that provides a new technical solution to do something (Farre et al., 2020).

Due to the exciting advances being presented daily in AR, there has been an imperative need to protect them. Thanks to the development of patents, it is expected that the evolution and consolidation of AR could occur faster. Figure 10.3 shows a graphic generated with lense.org regarding the worldwide AR number of patents granted from 2013 to 2023.

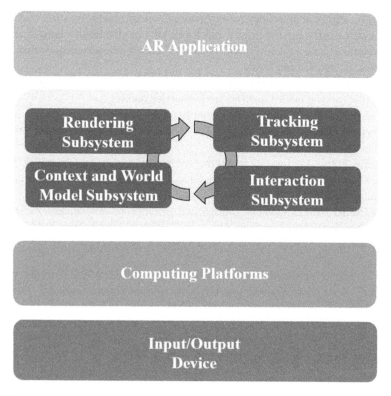

FIGURE 10.3   AR patents were granted from 2013–2023.

Figure 10.3 depicts a progressive growth in the number of patents granted in the last 10 years, showing interest in using AR in different fields and that it will soon enter its state of maturity. The number has gone from 1279 patents in 2013 to 12,814 until October 2023, representing an increase of 1000%. According to Evangelista et al. (2020), patents on AR are mainly oriented toward software (application design), hardware (capture and display devices), user interaction (using senses), and tracking. Moreover, most patent owners are companies such as Apple, Amazon, Meta, and Microsoft, and unfortunately, just a few patents pertain to research centers or universities.

According to Maddikunta et al. (2022), AR is one of the technologies expected to assist the next industrial revolution, called Industry 5.0. With the appearance of Apple Vision Pro glasses, AR users are contemplated to increase, and industries will be encouraged to insert the technology into manufacturing processes. Consequently, more patent applications will be filed and granted. Readers interested in increasing their knowledge of AR patents can consult the works of Choi et al. (2018), Evangelista et al. (2020), and Jeong and Yoon (2017).

10.6   LESSONS LEARNED

In 2011, the authors of this book directed the first thesis on AR, and since then, they have made many efforts to make scientific contributions to this fascinating field. Additionally, in 2014, they created one of Mexico's first computer vision and AR laboratories. Today, 12 years later, we are convinced that AR is a field that still has a lot to offer. However, providing knowledge in the area of AR is not easy, so it is considered appropriate to share some lessons learned.

Computer vision and programming are the main areas of knowledge in designing a fully usable AR application. Therefore, new developers should take introductory to advanced programming and computer vision courses. The 3D modeling area is also essential to design the virtual models that will be superimposed on the experience. However, if developers have a clear idea of what they need, they can ask experienced designers for help.

New developers should not become desperate or overwhelmed. Please remember that developing AR applications is still a rather technical subject. Therefore, it is recommended to start with the development of simple applications in which the basic functionality of AR can be tested. Once the function of computer vision libraries is understood and programming skills increase, more usable AR applications can be developed.

In AR, as in any science, it is crucial to stay informed of the advances that occur every day. Therefore, reading books and divulgation and scientific papers is recommended. Moreover, developers should attend technology expos and premier academic conferences like the International Symposium on Mixed and Augmented Reality (ISMAR) to explore commercial and research activities regarding AR. Also, it should be desirable to maintain attention on the development of standards and the publication of patents.

When the developer is asked to make an application, a deep and professional analysis must be conducted regarding the topic on which AR will be implemented.

It will be easier to contribute if the topic requires viewing objects from different perspectives. It has been proven that AR is not a technology that can be applied to any field. Poorly applied AR can cause stress and discomfort to users. In addition, the cognitive load when developing the task can increase.

Regarding education, several studies have shown the effectiveness of AR. However, it is still unclear which characteristics of the topics can be addressed with AR. The first time students try an AR experience, they get excited and want to keep using it. Unfortunately, there are no longitudinal studies that have shown the long-term benefits of AR in educational settings.

After the long journey of reading the 10 chapters of this book, all that remains is to thank the readers and ask them the following question: Are you willing to try the wonderful world of augmented reality?

REFERENCES

Arena, F., Collotta, M., Pau, G., & Termine, F. (2022). An Overview of Augmented Reality. *Computers, 11*(2), 1–15. https://doi.org/10.3390/computers11020028

Azuma, R. (1997). A Survey of Augmented Reality. *Presence: Teleoperators and Virtual Environments, 6*(4), 355–385. https://doi.org/10.1162/pres.1997.6.4.355

Barraza, R. (2015). *Arquitectura de Software para el Diseño de Material Didáctico Basado en Realidad Aumentada Móvil Colaborativa* [Ph.D. dissertation]. Universidad Autónoma de Ciudad Juárez.

Barraza, R., Vergara, O., & Cruz, V. (2015). A Mobile Augmented Reality Framework Based on Reusable Components. *IEEE Latin America Transactions, 13*(3), 713–720. https://doi.org/10.1109/TLA.2015.7069096

Bently, L., Sherman, B., Gangjee, D., & Johnson, P. (2022). *Intellectual Property Law* (1st ed.). Oxford University Press.

Buchner, J., & Kerres, M. (2023). Media Comparison Studies Dominate Comparative Research on Augmented Reality in Education. *Computers & Education, 195*, 1–12. https://doi.org/https://doi.org/10.1016/j.compedu.2022.104711

Caudell, T., & Mizell, D. (1992). Augmented Reality: An Application of Heads-Up Display Technology to Manual Manufacturing Processes. *Proceedings of the Twenty-Fifth Hawaii International Conference on System Sciences (HICSS)*, 659–698. https://doi.org/10.1109/HICSS.1992.183317

Chang, H., Binali, T., Liang, J., Chiou, G., Cheng, K., Lee, S., & Tsai, C. (2022). Ten Years of Augmented Reality in Education: A Meta-Analysis of (Quasi-) Experimental Studies to Investigate the Impact. *Computers & Education, 191*, 1–24. https://doi.org/https://doi.org/10.1016/j.compedu.2022.104641

Choi, H., Oh, S., Choi, S., & Yoon, J. (2018). Innovation Topic Analysis of Technology: The Case of Augmented Reality Patents. *IEEE Access, 6*, 16119–16137. https://doi.org/10.1109/ACCESS.2018.2807622

Dargan, S., Bansal, S., Kumar, M., Mittal, A., & Kumar, K. (2023). Augmented Reality: A Comprehensive Review. *Archives of Computational Methods in Engineering, 30*(2), 1057–1080. https://doi.org/10.1007/s11831-022-09831-7

Evangelista, A., Ardito, L., Boccaccio, A., Fiorentino, M., Messeni, A., & Uva, A. (2020). Unveiling the Technological Trends of Augmented Reality: A Patent Analysis. *Computers in Industry, 118*, 1–15. https://doi.org/https://doi.org/10.1016/j.compind.2020.103221

Farre, J., Hegde, D., & Ljungqvist, A. (2020). What Is a Patent Worth? Evidence from the U.S. Patent "Lottery." *Journal of Finance, 75*(2), 639–682. https://doi.org/10.1111/jofi.12867

Giovannucci, D., & Ponte, S. (2005). Standards as A New Form of Social Contract? Sustainability Initiatives in the Coffee Industry. *Food Policy*, *30*(3), 284–301. https://doi.org/https://doi.org/10.1016/j.foodpol.2005.05.007

Graham, S., Gillespie, A., & McKeown, D. (2013). Writing: Importance, Development, and Instruction. *Reading and Writing*, *26*(1), 1–15. https://doi.org/10.1007/s11145-012-9395-2

Jeong, B., & Yoon, J. (2017). Competitive Intelligence Analysis of Augmented Reality Technology Using Patent Information. *Sustainability*, *9*(4), 1–22. https://doi.org/10.3390/su9040497

Linowes, J., & Babilinski, K. (2017). *Augmented Reality for Developers: Build Practical Augmented Reality Applications with Unity, ARCore, ARKit, and Vuforia*. Packt Publishing Ltd.

López, J., Moreno, A., López, J., & Hinojo, F. (2023). Augmented Reality in Education. A Scientific Mapping in Web of Science. *Interactive Learning Environments*, *31*(4), 1860–1874. https://doi.org/10.1080/10494820.2020.1859546

MacIntyre, B., Gandy, M., Dow, S., & Bolter, J. D. (2005). DART: A Toolkit for Rapid Design Exploration of Augmented Reality Experiences. *ACM Transactions on Graphics*, *24*(3), 932–932. https://doi.org/10.1145/1073204.1073288

MacWilliams, A., Reicher, T., Klinker, G., & Bruegge. B. (2004). Design Patterns for Augmented Reality Systems. *Proceedings of the International Workshop Exploring the Design and Engineering of Mixed Reality Systems* (MIXER), 1–8.

Maddikunta, P., Pham, Q., Prabadevi, B., Deepa, N., Dev, K., Gadekallu, T., Ruby, R., & Liyanage, M. (2022). Industry 5.0: A Survey on Enabling Technologies and Potential Applications. *Journal of Industrial Information Integration*, *26*, 1–31. https://doi.org/https://doi.org/10.1016/j.jii.2021.100257

Mossel, A., Schönauer, C., Gerstweiler, G., & Kaufmann, H. (2012). ARTiFICe – Augmented Reality Framework for Distributed Collaboration. *International Journal of Virtual Reality*, *11*(3), 1–7. https://doi.org/10.20870/IJVR.2012.11.3.2845

Nadvi, K., & Waltring, F. (2004). Making Sense of Global Standards. In Schmitz, H. (ed.) *Local Enterprises in the Global Economy* (pp. 1–31). Edward Elgar Publishing. https://EconPapers.repec.org/RePEc:elg:eechap:2824_3

Perey, C., Engelke, T., & Reed, C. (2011). Current Status of Standards for Augmented Reality. In Alem, W. (Ed.), *Recent Trends of Mobile Collaborative Augmented Reality Systems* (pp. 21–38). Springer New York.

Schmalstieg, D., Fuhrmann, A., Hesina, G., Szalavári, Z., Encarnação, L., Gervautz, M., & Purgathofer, W. (2002). The Studierstube Augmented Reality Project. *Presence: Teleoperators and Virtual Environments*, *11*(1), 33–54. https://doi.org/10.1162/105474602317343640

Seichter, H., Looser, J., & Billinghurst, M. (2008). ComposAR: An Intuitive Tool for Authoring AR Applications. *Proceedings of the 7th IEEE/ACM International Symposium on Mixed and Augmented Reality (ISMAR)*, 177–178. https://doi.org/10.1109/ISMAR.2008.4637354

Simonetti, A., & Paredes, J. (2013). *Vuforia v1. 5 SDK: Analysis and Evaluation of Capabilities*. Universidad Politécnica de Cataluña.

Suppe, F. (1998). The Structure of a Scientific Paper. *Philosophy of Science*, *65*(3), 381–405.

Sutherland, I. (1968). A Head-Mounted Three Dimensional Display. *Proceedings of the Fall Joint Computer Conference, Part I*, 757–764. https://doi.org/10.1145/1476589.1476686

Yuan, Y. (2018). Paving the Road for Virtual and Augmented Reality [Standards]. *IEEE Consumer Electronics Magazine*, *7*(1), 117–128. https://doi.org/10.1109/MCE.2017.2755338

Index

Milton Keynes UK
Ingram Content Group UK Ltd.
UKHW031131141024
449569UK00006B/263

9 781032 563718